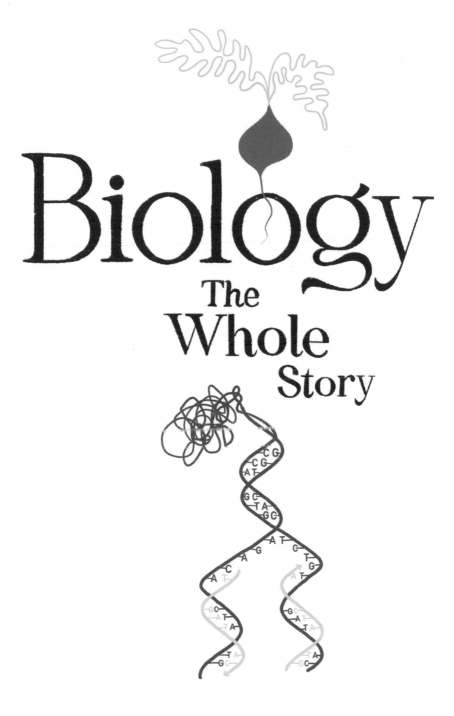

Biology
The
Whole
Story

www.davidficklingbooks.com

For Rowan

Biology - The Whole Story
is a
DAVID FICKLING BOOK
First published in Great Britain in 2023 by
David Fickling Books,
31 Beaumont Street,
Oxford, OX1 2NP
Text © Lindsay Turnbull, 2023
Illustrations © Cécile Girardin, 2023

978-1-78845-193-2
1 3 5 7 9 10 8 6 4 2

FSC
www.fsc.org
MIX
Paper from
responsible sources
FSC® C104723

DAVID FICKLING BOOKS Reg. No. 8340307
A CIP catalogue record for this book is available from the British Library.
Printed and bound in China by Toppan Leefung

Biology
The
Whole
Story

Lindsay Turnbull

Illustrated by Cécile Girardin

David Fickling Books

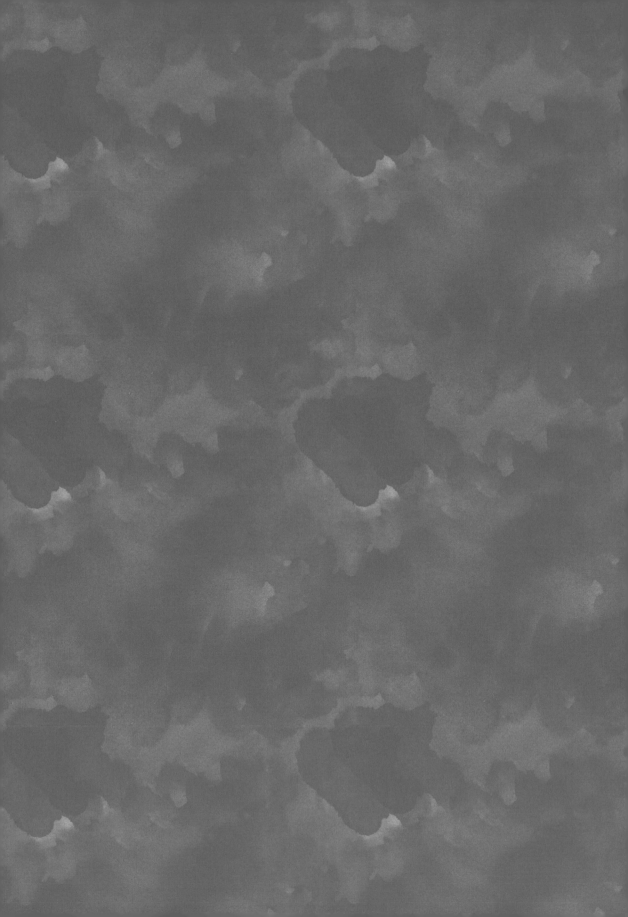

Contents

INTRODUCTION
Why this book?

Physics is often described as the science of the twentieth century, but biology has staked a bold claim to be the science of the twenty-first. Every day we are bombarded with new and important biological findings, from new vaccines for deadly viruses to strange new species lurking in hitherto unexplored places. Of course, the physicists haven't yet thrown in the towel. They continue to vie for public attention by enticing us with exciting new projects, and most recently they have turned their attentions to our neighbouring planet, Mars.

NASA's *Perseverance* rover is a robotic vehicle that landed on Mars in 2021 to trundle around its surface, collecting samples and setting them aside for future human visitors to inspect. The mission is a spectacular feat of human ingenuity and the scientists at NASA are rightly proud of their achievements. But the stated goal of the mission is unmistakeably biological – to look for signs of ancient past life on Mars's dusty surface – because today, its thin atmosphere and lack of liquid water mean that the red planet is undoubtedly a dead planet.

Personally, I'm happy to give Mars a miss. It's not that Mars isn't beautiful in its austere way, nor that it doesn't have some attractions, like the largest volcano in our solar system. But Mars, like so many of us humans, is utterly eclipsed by its gorgeous, stunning sibling. Why would you want to go to Mars when you can live on its shining sister – Earth?

Earth is unmistakeably a living planet. Suspended in the blackness of space, oceans of deep-blue water hold glowing green continents in their liquid embrace. The Earth's atmosphere is thick – and rich in oxygen gas – quite unlike the atmosphere of every other dull and lifeless rock that's scattered around our solar system. And it teems with unimaginable wonders.

Earth is currently home to at least eight million species of animals and plants – the products of four-and-a-half-billion years of evolution – and all earthlings have free access to at least part of this bounty. We can gasp at the aerobatics of a passing swift, or shudder at the scuttling arthropods that rush out from underneath the nearest rock. Even a drop of pond water contains a twirling, whirling frenzy of micro-organisms, which any microscope can reveal.

We ourselves are just one example of evolution's creative frenzy. Humans have undoubtedly been highly successful, spreading around the globe and reaching every continent. But we don't live quietly alongside other species – instead, we have transformed the planet, removing entire ecosystems and replacing them with others that more closely serve our needs. Indeed, so profound is our impact, that many now believe our planet may not support us for much longer.

Today, many children worry deeply about the planet-scale mess that they stand to inherit. If they are to tackle the challenges ahead, then a good grasp of biology is essential; but this is a daunting prospect. Biology is an enormous subject that grows every day, and the problems that biologists are expected to solve keep mounting: we need better antibiotics; we must prevent the emergence of the next pandemic; and all the time, species are disappearing, as the diversity of life on Earth comes under pressure from a growing human population that demands more every day.

I certainly sympathize with those children. I teach students at the University of Oxford, and they too worry about their future. They often feel overwhelmed by the immensity of biology, and it's my job to try to make it manageable. Studying the school curriculum and looking at other sources of readily available information led me to conclude that there are two intertwined problems with teaching and learning biology today. The first problem is that biology does, indeed, continue to grow at an alarming rate. Biologists begin with an idea and then further investigations add more knowledge and the original idea might start to fracture into several sub-ideas, each with their own supporting evidence. In the end, this continued growth makes it difficult to teach any biological topic, as it's extremely hard to know what is essential and what is simply nice to know.

The second problem is the tendency to serve up biology in rather large indigestible chunks. Even the best textbooks continue to lay biology out in a rather dull way and don't attempt to turn it into an enjoyable read. It's the student's job to simply learn the information, and there isn't any real context or story to help the reader along.

This is strange because the history of life on our planet is the most incredible story. It begins with a molten rock being bombarded by meteorites and it ends with what you see around you today: a teeming world of species that interact with each other and their surroundings. Between that distant hot rock and the present day, millions of living things have sprung into being, and although most have since vanished, we can learn something new from each and every one of them.

The story of life on Earth provides the perfect backdrop for the important biological concepts that we all need to know, and I have used that story to structure this book. You'll be glad to know that I don't give equal space to each period of the Earth's history – indeed this would be a recipe for a very dull book – as for very long periods we honestly don't really know what was happening. It also doesn't follow a slavish chronology – my goal is to tell the story in the way that makes it easiest to understand, so animals get two chapters before plants get to muscle in and tell their story.

Despite my best efforts, many episodes in the story of life are still somewhat controversial. There are also many things that we simply don't know. I believe that the best way to deal with this issue is to be honest, in the hope that this might inspire some readers to fill these gaps by becoming biologists themselves. So, I hope that you will find these gaps reassuring – there's still plenty for budding biologists to find out!

Finally, when writing about science, a decision has to be made about the scientists themselves, and mostly I decided to leave them out. I accept that science is done by real people and obviously they deserve appropriate credit, but mentioning them all means introducing yet more names to a subject that is already full to bursting with new terminology. To deal with this particular problem, there is a glossary at the back of this book and any technical word written in italics, like *cell nucleus*, can be found there, together with a brief definition. The book also contains illustrations. These have been developed in conjunction with my fantastic colleague, Cécile Girardin, and are simpler than those found in most text books. They are not a replacement for a text-book diagram and we have allowed ourselves a certain artistic licence when we think it helps to make a tricky concept clearer (a ribosome is not really a stout man in braces!).

In writing this book, I relied on the expertise of many incredible colleagues at the University of Oxford, who shared their knowledge with the relish and eagerness that typifies most biologists. Of course, I must take the blame for any errors. As I have already explained, new discoveries are made every day, but I still believe that many of the core concepts in this book have stood the test of time and are unlikely to change dramatically. So, if each biological topic can be thought of as a delicious dish, then future biologists might tinker slightly with the ingredients, but I doubt that they will entirely change the recipe.

So, now, let's begin with the fundamental unit of all life on Earth – the cell. These tiny entities form the building blocks of larger creatures, but most of the cells on Earth live alone – quietly passing on their information to the next generation and turning long-evolved plans for world domination into action. Cells might be small but don't let their size fool you. These tiny titans are the ultimate life form and if cell biology were a delicious dish, then it would be the equivalent of the best cake you've ever eaten.

Chapter 1
INFORMATION
In the beginning was the word

Every life form on our planet is made from *cells*. They are the fundamental unit of life, and for most of Earth's history, cells lived alone and did not join forces to build larger beings. But today, our planet is populated by bodies of all shapes and sizes. Each one is a collaboration between enormous numbers of cells that work together seamlessly, so it's unfortunate when their collective efforts are undermined by cells that don't properly play their part.

Each morning, Sara sits in the window of her house and watches the world go by, and like most of us, she is oblivious to the internal workings of her body. But Sara's world is a difficult one. She suffers from a condition called *cystic fibrosis* and must endure hours of physiotherapy to clear deep-seated mucus from her lungs. Yet, even with this help, she has spent long periods of her short life in hospital, receiving treatment for bacterial infections that threaten her very existence.

Inside Sara's lungs, we can begin to find out what has gone wrong. Thick sticky mucus has accumulated in the tiny air passages, allowing bacteria to thrive and so making Sara ill. This problem has arisen because some of the cells within her body are malfunctioning. A human body is made from around 37 trillion cells, and most of us are lucky enough that those cells carry out their assigned functions most of the time. So, why are some of Sara's cells causing her so much harm?

To answer this question, we need to understand how cells know what it is they are supposed to be doing. Studying cells means peering into their interiors, but cells are so small that they can't generally be seen with the naked eye. Fortunately, since the invention of the microscope, scientists have gradually found out more and more about them, revealing orderly interiors sculpted from even tinier molecules. Their minuscule size can deceive us into thinking that individually they are dull and somewhat lifeless – perhaps similar to a tiny Lego brick – but nothing could be further from the truth.

Far below our range of vision is a world of mind-boggling complexity. A cell is a tiny hive of ceaseless activity – more akin to Willy Wonka's chocolate factory than a lifeless piece of plastic. If we could win a golden ticket to visit this factory, we would see workers hammering out new tools and spare parts at an extraordinary rate and being bombarded by messages flying out from head office with instructions to make and repair essential components; and if we stayed long enough, we might witness the most momentous time in the life of any cell – when it pulls itself apart to become two identical descendants.

Many of us learn about cells for the first time at school, and it's common to draw pictures of 'typical' animal and plant cells. These cells are generally large, and the diagrams highlight a few prominent features that are visible under a light microscope: an outer *cell membrane* that separates the cell's contents from the world outside; a *cell nucleus* that acts like a head office; and a watery *cytoplasm* that acts like the factory floor, where the business of the cell is mostly carried out. However, there are many other types of cells on our planet, including bacterial cells that far outnumber animals and plants, and these cells look deceptively simpler – for example, they don't have a cell nucleus – which might delude us into thinking that they are fundamentally different. But if we focus on what cells *do*, rather than on what cells look like, then we might draw a different type of diagram to highlight the handful

Fundamentals of a Cell

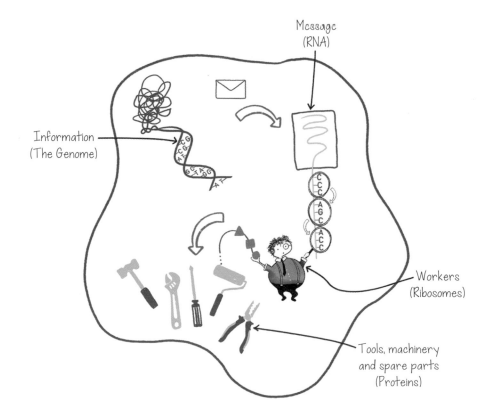

Message
(RNA)

Information
(The Genome)

Workers
(Ribosomes)

Tools, machinery
and spare parts
(Proteins)

of crucial features that all cells share, and most of these are central to processing the information that lies at the heart of every cell.

All cells, whether animal, plant or bacteria, have to follow instructions to ensure that the cell runs smoothly, and their instruction manual is called a *genome*. A genome is generally too small to see, even with a light microscope, but that doesn't make it any less valuable. The importance of carefully written instructions is clear to many of us when we buy things like flat-pack furniture that require some assembly. If the finished object is to look like the one shown on the box, then the pieces must be correctly assembled, and the instructions show the buyer how to do this. In a cell, the genome acts in the same way – providing precise instructions for constructing everything that the cell needs using tiny building blocks that the cell either imports or makes from scratch.

Of course, someone needs to read the instructions and build the finished products. To perform this role, cells are filled with miniaturized workers called *ribosomes*, and if we compare the cell to Willy Wonka's chocolate factory, then these are the Oompa-Loompas. Ribosomes are inexplicably missing from many of the diagrams drawn at school, but cells would grind to a halt without these indefatigable workers.

Surprisingly, perhaps, ribosomes aren't allowed direct access to the genome. Instead, messages are sent out that specify precisely what needs to be made. Each cell might make hundreds or thousands of different pieces of equipment and the ribosomes are capable of making any of these, as long as they receive the right instructions. And only when it has the right equipment can the cell perform its proper function.

So, the genome provides detailed instructions to the ribosomes, and they build all of the machinery that the cell needs. This transfer of information from genome to ribosomes is central to the operation of all cells, but they don't just process information so that they can run their own lives, cells also pass information on to the next generation. Sara, like every multicellular being on Earth, began her life as a single cell. To create the 37 trillion cells that she now consists of, this first cell had to divide, again and again, producing vast numbers of descendants to take up every role in Sara's body. But before a cell divides, it must copy the all-important genome, so that each of its descendants, often called daughter cells, can properly manage their own affairs.

These two acts – passing on information to the next generation of cells and translating that information into action – are the most fundamental things that cells do, and they are carried out by every cell on Earth. This information processing is so important that we need to understand it in more detail. Only by doing so can we get to grips with how organisms truly work, how things go wrong and, crucially, how the first cells on our planet came into being.

Written in a simplified language that no human would recognize, the genome contains the instructions for building every machine and spare part that the cell could possibly need. Laid out on an enormous scroll, the genome is kept by different cells in different places. In bacteria it simply floats around, while in the larger cells, like those from which we and other complex beings are made, the genome is kept apart within the cell nucleus, although its function is the same.

The instructions within the genome are spelled out using DNA, an *information molecule*, whose sole purpose is to store information and keep it safe. Like most of the large molecules inside cells, long strands of DNA are formed by joining together smaller building blocks just as we could make a necklace by stringing together glass beads. The building blocks of DNA are called *nucleotides* and there are four of them: adenine, cytosine, guanine and thymine. They are denoted by their four initial letters – A, C, G and T – which are used to spell out each instruction. Different organisms need different instructions, depending on whether or not they have legs, wings or leaves, and so the genome of each organism is unique.

We can begin to appreciate the enormous complexity of living things by noting that the genomes possessed by even the simplest organisms are massive. Building and running a bacterium requires a genome that clocks in at around one million DNA letters, a worm needs one hundred times that, while the genomes of more complex creatures, like sharks, pigeons or horses, contain at least one billion letters and usually more. Indeed, a genome contains so much information that only

The Genome is Written in DNA Letters

a miracle of coiling and packaging allows it to squeeze inside the cell.

Instructions are sent out from the genome by a second information molecule called *RNA*. Strands of RNA are made by stringing together building blocks that are very similar to the ones used to construct DNA strands, but they have their own nucleotide alphabet: A, C, G and U: adenine, cytosine, guanine and uracil. So, given the similarity between the two information molecules, we need to understand why the cell needs both.

RNA strands are short, consisting of a few thousand letters at most. They are perfect for delivering simple instructions, but wouldn't be much use for storing the billions of letters needed to encode the genome of a horse or a shark. But, DNA has a trick up its sleeve. A strand of DNA letters is always bonded to a second, complementary strand and the two coil around each other to form a structure rather like a spiral staircase, with the letters hidden deep inside. Known as a double helix, this arrangement is very robust and means that the cell can keep adding more and more letters to the DNA molecule without it falling apart.

The double helix of DNA allows enormous genomes containing billions of letters to remain intact for the hours, days, weeks, months or even years that some cells survive. Indeed, without this miracle-molecule, complex animals and plants with massive genomes simply couldn't have evolved. Once the cell dies, DNA slowly degrades, but it's stable enough to allow us to read the sequence of letters inside the dried-up genomes of long-dead plants, animals and even people – so by digging up mummies from their tombs, we can learn more about ancient Egyptians, but we won't be able to resurrect dinosaurs (whatever the makers of *Jurassic Park* may claim).

While the cell is alive, DNA can't just sit around in splendid isolation – if the cell is going to make things, then its instructions must be put to use. If the cell needs a particular piece of machinery or a spare part, then the double helix is prised apart. An individual instruction within the genome is called a *gene*, and by exposing the sequence of DNA letters for the gene required, RNA messengers can faithfully copy the gene's code into their own alphabet and carry it away. But only the industrious ribosomes know how to decode the message and turn it into a piece of viable kit.

Most of the objects that humans knowingly interact with are large, with sizes ranging from a few centimetres to a few metres, so cells are too small for most of us to notice. An *amoeba* is about the largest free-living cell out there and 10 of them lined up side by side would fit into one millimetre, while for the typical cells inside a human body, around fifty to one hundred of them could probably fit into the same space. Bacterial cells are smaller again, so fitting one thousand or so into our millimetre wouldn't be too hard. But whichever cell we choose, its internal parts and machinery are very small indeed, which has dramatic effects on how things work.

The world we inhabit – filled with objects measured in centimetres and metres – is dominated by gravity. Earth is massive, so it attracts other large objects and keeps them rooted to one spot. Gravity dictates that our factories are laid out on a flat surface, where machines stay in place and people walk around – hence the factory floor – but the components of a cell are far too small to feel gravity's pull.

Instead, the inside of a cell is more like a factory inside the international space station, where everything is weightless and things float around. Within the crowded cell, tiny molecules often bump into each other, and while most of these collisions simply result in them bouncing off each other again, if two molecules are the right shape, then they will stick together and something more interesting might happen.

The RNA messages floating out of the genome are hoping to bump into a ribosome – one of the cell's key workers. Their job is to capture the RNA messages and use the information to build something useful, and with up to 10 million of them within a single cell, bumping into one shouldn't be too hard. In terms of scale, they are to the cell what individual people are to a city like London. A ribosome is so tiny that no regular microscope will ever reveal one, but more advanced techniques have shown that, unlike city dwellers, they resemble tiny snowmen with fat bodies and large heads, and the gap between body and head is just the right size and shape to trap the RNA message.

The job of a ribosome is to build *proteins*. Proteins play a multitude of roles within cells. Some are structural – forming scaffolds to support the cell – while others act as signals so the cell can communicate with the world beyond its outer membrane; but the biggest class of proteins are *enzymes* – molecular machines that form the equivalent of the cell's toolbox. Advanced microscopy has revealed that proteins can fashion an extraordinary array of tools and machinery, including tiny locks with perfectly fitting keys, channels through the membrane that only allow selected traffic to enter, and even the blades of molecular turbines that allow the cell to generate the energy it needs. To play all these roles effectively, proteins have to be built in a bewildering variety of shapes and sizes. But how does a single type of molecule manage such extraordinary flexibility?

Like the information molecules, a protein is made by sticking together smaller building blocks to form a long chain, but this time the building blocks are *amino acids* rather than nucleotides. The flexibility of proteins lies in the amino acids, because while the four letters in a DNA strand are rather similar, there are 20 possible amino acids to choose from, each with unique chemical and physical properties. Amino acids carry different electrical charges (positive, negative or neutral), bond to different substances, and some, like valine, hate water, while others, like arginine, love it.

Once the amino acids are strung together, the chain seems to take on something of a life of its own. The water-hating amino acids pull themselves towards the middle of the protein, while the water-loving ones are happy to be on the outside. Meanwhile, amino acids in different parts of the chain are repelled or attracted to each other by their varying electrical charges, and all this jostling means that the chains can fold and twist themselves into practically any shape we can think of. And because each protein contains a different sequence of amino acids, the shape of each protein is unique.

The shape of the protein is determined by the sequence of amino acids, so the ribosomes have to 'know' exactly which amino acids to join together, as otherwise the key won't fit the lock, the wrong molecules will pile up in the channel, and the turbine blades won't spin. But they don't have to trust to luck. This crucial information is encoded in the genome and delivered by the RNA messengers.

Protein Folding

 Water loving amino acid

 Water hating amino acid

Amino acids are joined together into a chain

The chain spontaneously folds into a 3-D shape. Here, the water–hating amino acids huddle together on the inside, while the water–loving ones are happy to be on the outside

Once trapped by a ribosome, the message simply has to be translated.

The messages flying out from the genome are written in RNA-speak, an unusual language in which all words are exactly three letters long. Each three-letter word specifies a single amino acid, so correct translation allows the ribosome to join amino acids together in exactly the right sequence. The ribosome begins by locking onto the message and finding the three-letter word AUG, which means 'start'. It then keeps moving – three letters at a time – to discover which of the 20 different amino acids comes next. If the next three letters are GUA, then the ribosome adds an amino acid called valine, but if the next three letters are GGA, then it adds an amino acid called glycine instead.

Ribosomes can find the right amino acid because each one is attached to a three-letter RNA tag. The ribosome just has to make sure that the tag on the amino acid complements the three-letter word on the RNA message – and if it does, then it's the right one to add next. Any number of amino acids can be added to the chain, although a typical protein in a human body contains between 350 and 400, but eventually, the ribosome encounters a word meaning 'stop', at which point it will detach itself from the message and the protein chain floats free. On release, the finished protein spontaneously folds into its three-dimensional shape, and if the sequence was right, then this shape allows the protein to perform its proper function.

One of the most astonishing revelations of twentieth-century biology is that RNA-speak is a truly universal language, understood by all. Whether the cells are embedded in the legs of a galloping horse or basking in the petals of an unfolding sunflower, they all know that GGA spells glycine – and it's not just animal and plant cells. In the heart of deep-sea vents, billions of bacterial cells living inside the guts of giant tube-worms are united in their certainty that GGA spells glycine.

RNA-speak first emerged around 3.8 billion years ago and many of its speakers haven't spoken to each other since, but the three-letter words for all 20 amino acids are written in exactly the same way – with almost no exceptions – in every cell on Earth. Contrast this uniformity with human language, which emerged around 100,000 years ago and has already diverged into at least 6,500 modern forms. People can

A Ribosome at Work

A ribosome must connect amino acids together in the right sequence

Message
(RNA)

Each three-letter
word corresponds
to one amino acid

Proline

Glycine

Serine

If the sequence is right,
the finished protein will
be the right shape

change their language quite easily because the consequences of doing so are unlikely to be severe, but if a cell tried to change RNA-speak, then every protein would be affected – and it's spectacularly unlikely that at least one of those changes wouldn't bring disaster. So, despite trillions of opportunities to evolve a new language, no cell has ever done so.

The one-way flow of information within cells – from genome via RNA messenger to protein – is called the *central dogma* of molecular biology. It is a universal system adopted by all cellular life and is one way to define what cells are: they are factories that process information and use it to build 'stuff' (formally called matter). But cells do more than just process information during their own lifetimes – they also pass information on to their descendants, and to do this, the genome must be copied.

The life of any cell is finite, and eventually it will either die or pull itself apart to produce two identical descendants. For single-celled organisms, the two daughter cells then separate and lead independent lives, but in complex beings, like humans or horses, the two cells stay together and keep dividing again and again. These divisions generate the enormous numbers of cells that are needed to fashion a complex being: a small leaf captures sunlight with around 100,000 cells; a human heart beats with the aid of two billion; and an adult human is kept walking and talking by 37 trillion. But all these ambitious building projects started in the same way – with a single cell that divided in two.

Producing two cells out of one is the sort of everyday miracle that life performs so effortlessly, we barely notice how remarkable it is. But before the cell divides, the genome must be copied, as otherwise the daughter cells will be lost and lifeless. To reveal the hidden letters, the DNA helix is untwisted and the two strands prised apart, which are then used as templates to build an identical copy using molecular copying machines that are built from enzymes.

Once pulled apart, each of the two strands can be used as a template to recreate the complementary strand that is now missing. This works smoothly because each

DNA letter will only pair up with one of the four letters on offer. For example, the letter A will only pair up with the letter T, so when the copying machine meets the letter A, it knows that a letter T is required on the new complementary strand that it is building. In turn, T will only pair with A, C will only pair with G and G with C. So, by working away and adding new letters, the copying machine produces two new identical double-stranded DNA molecules.

When faced with a three-billion-letter genome, a single copying machine in a typical animal cell would need around eight hundred hours to finish the job if it started at the beginning and continued all the way to the end. To avoid such an impractical wait, an army of machines starts in multiple different places at the same time, and so can finish copying a three-billion-letter genome in around one hour – about the time it would take 1,000 crack typists to copy out the complete works of Shakespeare.

DNA Replication

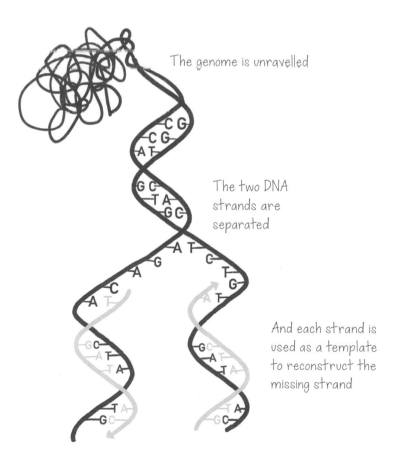

The genome is unravelled

The two DNA strands are separated

And each strand is used as a template to reconstruct the missing strand

Before a **human** cell divides, **two** metres of DNA must be copied

Inside multicellular beings, the copying process doesn't stop once adult size is reached. Within bodies, cells need to be continually replaced and each new cell needs a copy of the genome. In complex organisms, like humans, the amount of copying needed to generate enough cells for a lifetime is truly astronomical. By the time we reach 40 years old, it's estimated that our bodies have produced a light-year of DNA (9 trillion kilometres) – enough to extend well beyond the outer reaches of our solar system, although not quite enough to get us to the next star.

A genome can be copied again and again, giving it a life span far beyond that of individual cells or organisms. Indeed, our bodies are just disposable shells and easily discarded. Only information is truly immortal, and because new cells can only arise from old ones, the transmission of information between generations is crucial to life as we know it.

The molecular machines that copy DNA are fantastically good at their job, so the genome handed to every daughter cell should be identical. But in truth, the copying machinery isn't perfect. Although it has built-in proofreading, it sometimes makes mistakes, and these mistakes can have devastating consequences for the daughter cells that inherit them.

It's time to return to Sara and examine the cause of her illness. Sara suffers from cystic fibrosis, which mainly affects the lungs, but how do Sara's lungs differ from those of non-sufferers?

In non-sufferers, the cells that line the air passages in the lungs make a very large protein, called the CFTR protein. This is inserted into the cell's outer membrane where it forms a channel allowing charged particles, called ions, to enter and leave the cell. With ions flowing freely, the cells are able to secrete thin mucus that performs its intended function of trapping any bacteria that enter the airways, which are then swept away by the tiny hairs that cover the cell's surface.

In Sara's lungs, these channels appear to be missing. Without them, ions can't flow freely in and out of her cells, so they can't secrete mucus properly either. The thick sticky mucus – which is all her cells manage to produce – also traps bacteria, but it can't be swept away, leaving the bacteria to build up to dangerous levels. So, why are these crucial channels missing?

Zooming inside her cells, we see that messages to build the CFTR protein are certainly flying out of her genome and being trapped and translated by ribosomes. But something must have gone wrong because the CFTR protein they are making is one amino acid short. It should contain 1480 amino acids, but, in Sara's cells, the 508th amino acid is missing, and this is catastrophic for the protein's final shape – it simply won't form an acceptable channel, and so it is rejected and its amino acid parts recycled.

The fault does not appear to lie with the ribosomes, as they are successfully making other proteins, so something must be wrong with the message itself. In fact, the message sent to the ribosomes is too short because three crucial letters are missing from the relevant part of Sara's genome. At some point, three tiny letters were accidentally deleted from the genome of the founding cell of Sara's body, and because all the cells in her body are clones, they have all inherited the same mistake.

Although Sara has the benefit of various drugs to ease her symptoms, a cure would require editing the genome within each of her cells to insert the missing letters in exactly the right place. Such technology, called *gene editing*, is developing rapidly, and it may be that sufferers of cystic fibrosis in the future can benefit from totally new kinds of genetic treatment. But until this technology is perfected, Sara must rely on more traditional methods to manage her condition.

Cystic fibrosis is one example of a genetic disorder. Humans suffer from thousands of genetic disorders, some incredibly rare and others more common, and while some have only mild effects, others can be life-changing. Regardless of their severity, all genetic disorders have the same cause: a change to the genome that alters one of the RNA messages sent out to the ribosomes. Of course, it might be better if the ribosomes noticed the mistake, but unfortunately, in contrast to the dedicated Oompa-Loompas who worked in Willy Wonka's chocolate factory, ribosomes are nothing more than mindless machines. On receipt of a faulty message, ribosomes don't ask questions, but churn out the altered protein, unaware of any potential problems. The consequences range from mild to serious to utterly catastrophic, and there's absolutely nothing that the cell – or the body to which it belongs – can do about it. The change to the genome that causes cystic fibrosis is an example of a *mutation*, where DNA letters are missing, added or simply swapped around. In films, mutations are usually caused by mad scientists and experiments in their laboratories, and there are certainly some chemicals, called *mutagens*, that damage DNA. But the vast majority of mutations have a more humdrum origin: they are simply mistakes made by the cell's own copying machinery.

Mutations in the genome alter the RNA messages flying out to the ribosomes, and the bigger the mutation, the more likely it is to have dramatic effects. But the size of the change isn't the only important factor. The unique nature of RNA-speak means that even swapping one DNA letter for another can have very different impacts on the cell – and the body to which it belongs – depending on exactly *where* the change takes place.

Each three-letter word in the RNA message spells out a single amino acid and all cells use the same language. The three-letter word GGA spells out glycine, but glycine has three alternative spellings: GGC, GGG and GGU – all equally good – and the three alternatives are also recognized by all cells everywhere.

Alternative spellings of the same amino acid mean that some mutations are *neutral*. If GGA mutates into GGC then the protein chain remains exactly the same, because both words spell out glycine; but if GGA mutates into GCA then the meaning of the word has changed and glycine will be replaced by a different amino acid – in this case alanine. Even so, the change might not affect the function of the protein, if its shape stays the same, but swapping just a single amino acid can cause a change in shape large enough to give rise to problems.

For some people, a sudden burst of exercise can leave them in severe pain for days or even weeks. The pain is caused by misshapen red blood cells getting stuck in the tiny vessels that deliver oxygen to the farthest reaches of the body. Normal red blood cells are squishy, allowing them to squeeze easily through narrow spaces, but sufferers of a genetic disorder called *sickle cell disease* build red blood cells that are just too rigid.

Sickle cell disease is caused, not by the loss of letters, but by a change to a single letter of the genome within the instruction (or gene) to make an oxygen-carrying protein called *haemoglobin*. Red blood cells are stuffed full of haemoglobin, and if the protein is the wrong shape, then the cells are too. The mutation changes the three-letter word GAG to GTG, and when the ribosomes encounter this word during the building of the protein, they use the amino acid valine instead of the usual glutamic acid. This single change is enough to transform the shape of the protein and with it the crucial red blood cells on which our bodies depend.

The central dogma means that all cells are slaves to their genomes. They can only process the information they have been given – there is no way for cells to rewrite their genomes or for ribosomes to leave feedback on the quality of the proteins they build. Mindlessly engaged in turning written instructions into three-dimensional objects, ribosomes are wonderful machines – but if the instructions are faulty, they aren't interested in our complaints.

The Effect of Mutation

▲ Water loving amino acid

■ Water hating amino acid

A mutation has altered a single amino acid in the chain

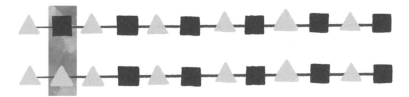

The effect is to change the shape of the protein

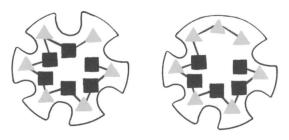

This is likely to mean that it no longer functions properly

Harmful mutations might be so severe that a cell or body is lost before its genome can be passed on, but for those cells or bodies that manage to reproduce, information flows forward into the next generation. So, if each new cell requires an existing cell to bring it into being, then all cells must have an unbroken line of ancestors that stretch back to the dawn of life on Earth. But if we follow different lines back through time, do they converge on the same ancestor?

The cell defines life on Earth and the defining features of cells reveal something undeniable about their origins. Cells share an extraordinary amount in common: they have a genome encoded by DNA using the same four letters; they send out instructions via RNA messengers that are written in RNA-speak; and these are decoded by near-identical machines called ribosomes that all speak the same language. Only one conclusion is possible – all life on Earth today is related, so there must be a founding cell from which we can all claim descent. Although there may have been other types of life in the past, they are not here now and we have no evidence for their existence. But amazing though this seems, it raises an even more mind-boggling question: we might know how a new life starts on Earth today, but how on earth did life itself get started?

Attempts to unravel the mysteries of life's origin began in the 1950s by focusing on the simple building blocks, like amino acids, from which larger molecules, like proteins, are made. Animal cells rely on their diet to obtain most of the building blocks they need, but the very first cells had to be much more self-sufficient. The early Earth only contained very simple molecules, like carbon dioxide and small amounts of methane, a natural gas, which we now burn to generate heat. So, how can cellular life have started from such unpromising beginnings?

The first serious attempt to solve this mystery took place in 1953 when a mixture of the very simple molecules thought to be present on the early Earth were repeatedly zapped with electricity to imitate lightning in the early Earth's atmosphere. To the world's astonishment, this unlikely attempt at primeval cookery yielded several different amino acids within the first week, so it seemed that it was easy to

In sub-Saharan Africa, **3%** of **babies** are born with sickle cell disease

conjure up the ingredients for life. But, hopes were dashed when it turned out that the first guesses at the chemistry of the early Earth might have been some way off the mark, and attempts to repeat the experiment under more realistic conditions were rather less successful.

Just when it seemed that the building blocks of life might resist all attempts at earthly conjuring, the cosmos came to our rescue. The early Earth was bombarded with meteorites and these are often loaded with the kind of building blocks that life needs – so their arrival might have kick-started life on our planet. Of course, this makes it quite likely that other planets in other systems have also received the same starter culture, so perhaps there's good reason to hope – or fear – that other life is out there.

Others believe that meteorites are a red herring – arguing that any extra-terrestrial building blocks would quickly have been used up – so is there somewhere on Earth where building blocks are continually manufactured? Perhaps the answer is not in space but closer to home, in the depths of the oceans. In 1977 marine scientists discovered strange new habitats on the ocean floor where seawater was interacting with molten rock deep within the crust in volcanically active areas. This interaction produces *deep-sea vents*, where hot chemically rich fluid bubbles up, forming black or white columns of streaming chemicals in the water, called 'smokers', that teem with life. These chemical brews are now prime contenders for the site of life's origin as they could be a natural source of amino acids – a key building block for early cells. But, while fascinating, the source of the building blocks isn't the greatest challenge faced by those who want to understand the origins of life.

To work out how the first cells arose, we need to solve the puzzle of the central dogma, which goes something like this. A cell relies on machinery built from proteins to carry out essential tasks – but it can't build these proteins without instructions. The instructions are provided by the information molecules – DNA, which stores the instructions and RNA, which carries them to the ribosomes for translation – but the instructions are useless without the machines to put these instructions into action. This tightly woven sequence leaves us with a classic chicken and egg conundrum, so which came first: the message or the machine?

To unravel the origins of the cell's deeply intertwined core biology, a bunch of assorted theories, some wild, some plausible, have been advanced over the decades. None has yielded an entirely satisfactory answer, and in 2019 a $10 million prize-fund was founded by the Royal Society in London for anyone who could finally solve it. The origin of the first cell may forever be shrouded in mystery – after all, it happened so long ago that it's rather like peering down the wrong end of a deeply cracked and clouded telescope at a star buried in a galaxy thousands of light years away; but that hasn't stopped us getting gradually closer.

It seems likely that the central dogma, which assigns different roles to different molecules, probably evolved from a simpler system where one molecule played all roles, but which one? DNA is chemically lifeless and it's impossible to imagine it doing anything other than simply quietly storing information. But RNA is quite different.

One of the shared features of all cells are the indefatigable ribosomes. These unsung heroes are unique among the cell's machinery in being constructed from both RNA and protein. Indeed, the active part of the ribosome – which can join amino acids together – is made from RNA, whereas nearly all other chemical activity within cells is carried out by enzymes (made from protein, not RNA).

The ability of RNA to carry out chemical reactions, while simultaneously storing information, gave rise to the idea that there was once an 'RNA world' in which RNA multi-tasked. Indeed, short strands of RNA can copy themselves without the help of enzymes, allowing information to be passed on. The theory goes that once a self-sustaining RNA world had developed, DNA took over the role of information storage and proteins became the masters of chemical activity – leaving RNA relegated to its current roles as ribosome and messenger.

Supporters of the RNA world point to one last remaining scrap of evidence. In 2020 a new disease called Covid swept the world, causing widespread lockdowns and disruption. The disease was caused by a *virus* – an agent that arose from living things – but like the black riders that galloped out of Mordor in *The Lord of the Rings*, is itself neither living nor dead.

Viruses are simply genomes wrapped in a protective layer. They contain information to build themselves but not the cellular machinery to turn those instructions into

A Virus

A virus injects its genome into the host cell

The viral genome sends out RNA messages

The ribosomes read the RNA messages and build viral proteins instead of tools for the cell

anything useful. Alone, they are nothing, and this impotence exposes why the message and the machine are so intimately connected in all cellular life. But viruses don't need their own machinery because they can enter cells and hijack their ribosomes, causing them to churn out viral proteins, including enzymes that copy viral genomes and package them up into new virus particles. Eventually, the cell bursts, releasing hundreds of new viruses into the world, each of which can go on to infect a new cell.

Viral genomes are intriguing because not all are made from DNA. Indeed, the virus that causes Covid has an RNA genome, and can perhaps trace its ancestry all the way back to the RNA world. If true, it means that viruses were present at the dawn of life, and perhaps even played a key role in getting cellular life up and running. Cells can only transmit information to their descendants, but by penetrating cells and moving information molecules around, viruses might have unwittingly swapped pieces of genome between different types of cells, and so produced new and exciting combinations.

The idea of an RNA world certainly holds many attractions, but not all scientists are convinced. Despite herculean efforts, longer strands of RNA have never been found to successfully copy themselves without the help of enzymes. Other scientists are working on theories that bypass the RNA world entirely. Perhaps life's origins will never be entirely clear and this is what makes it so fascinating. Competing theories go in and out of fashion and scientists fight to defend their favourites. What *is* clear is that much more remains to be done and that new evidence will continue to trickle out, as the cracked and clouded telescope gradually reveals long-hidden secrets from the dawn of life.

Tiny fossils from ancient hydrothermal vents reveal that the first cells were probably up and running by around 3.8 billion years ago, and they eventually gave rise to all the living things that have ever walked, crawled or rooted themselves upon our planet. But how do we get from single cells to prancing ponies and towering trees?

All life is defined by the information it carries, so this must have changed over the millennia. Mutations can generate difference, but most, like Sara's, are harmful, as cells and organisms are well-oiled machines and random changes are unlikely to be of any benefit. But just occasionally, a change to the genome results in a protein that functions rather better than the old one. Perhaps the key is now a slightly better fit to the lock or the turbine blades spin more smoothly – and such a change makes the cell or organism even better attuned to the world around it. But mutations occur in individuals, so how does a change to one end up being adopted by all?

To see how mutations can shape entire species, we need to look at an extraordinary force – one that can supersize dinosaurs or allow birds to take flight. *Natural selection* is a somewhat dull term for the ultimate power on our planet – one that elevates some and spells doom for others. But it's sometimes surprising who the winners and losers are.

Chapter 2
EVOLUTION
Misfits shall inherit the Earth

In 1819 in a woodland in Northern England, a moth shrugs off the shackles of its underground cocoon and crawls up the trunk of the nearest tree. Called a peppered moth, it has spent the winter sheltering below ground and must now extend two pairs of newly formed wings in preparation for its first flight. In this respect, as in nearly all others, it is no different from the other peppered moths emerging in the woods around it – but there is one crucial exception. This moth is a mutant, and instead of owning pale speckled wings like almost all the other members of its species, it has beautiful sooty-black ones. Given that most mutations are harmful, we might prophesy doom for this new arrival, but in the decades to come, naturalists will watch in awe as this moth and its descendants become wildly successful. But how?

At the start of the nineteenth century, the *Industrial Revolution* in Britain was well underway. Victorian engineers had worked out how to burn coal to heat water to produce steam to drive newly invented machines, so they could produce things cheaper and faster than ever before. Industrial development was greatest in the north of England and coal-powered industries flourished there. But coal is a dirty fuel.

Burning coal produces clouds of black soot, and the tiny particles settle out onto factories, homes, lungs and even trees. Worse, coal contains impurities, like sulphur – and when sulphur burns, it forms the acid gas sulphur dioxide – highly toxic to many of the crusty-looking grey-green *lichens* (pronounced: *like-uns*) that grow on tree trunks. In the space of a few decades, this deadly combination of soot and sulphur dioxide transformed the trees near English industrial towns from columns of speckled grey-green into pillars of purest black.

For a pale peppered moth sitting on a tree in nineteenth-century industrial England this transformation was disastrous. For thousands of years, the speckled wings of the pale peppered moths had allowed them to perch unseen on the light-coloured bark and encrusting lichens, but on the newly blackened tree trunks, the pale speckled moths stuck out like sore thumbs. The birds of Northern England did not let this opportunity go to waste. Moths make a good meal – and if birds can see them, they will eat them. So, was the peppered moth at risk of extinction?

Salvation lay in the mutant moth, whose sooty-black wings provided near-perfect camouflage against the newly converted tree trunks. Passing birds barely gave the

The Peppered Moth

black moth a second glance, so it was much more likely to survive and lay eggs than its *offspring* inherited the black wings and so, before long, the woods of Northern England were filled with black peppered moths, while the original pale moths were almost impossible to find. It might have been a rarity in 1819, but by the middle of the twentieth century, the black moth had become the new normal.

The story of the peppered moth reveals that species are not set in stone – they can *evolve* as the typical genome changes over time. Change begins with a lucky mutation in a single individual but it spreads through a group because those that carry it are more successful and have more offspring than those that don't. This almost ludicrously simple process might seem an unlikely way to generate the incredible diversity of life that we see around us, but it forms the core of the most important theory in biology: the theory of *evolution* by *natural selection*. Proposed in 1858 by Charles Darwin and Alfred Russel Wallace, who had independently hit upon the same idea, this theory has been the subject of frequent attacks by religious figures, and yet its central position at the core of biology has never been in serious doubt.

Naturalists during the nineteenth century weren't unduly troubled by the polluting machines of the Industrial Revolution. Instead, they were busy trying to answer a puzzle that had been bugging them for decades. The creatures that fill our planet appear to be spectacularly well-designed – geckoes stroll across the ceiling with gravity-defying ease, while the precision flying of hummingbirds allows them to sip nectar from the tiniest of flowers – but how did they end up this way?

Since the dawn of human history, the dominant explanations for well-designed creatures have always required a supernatural creator – and challenging this explanation has landed many people in serious trouble with religious authorities. But during the nineteenth century, people were seeking a more earthbound explanation for what became known as the Problem of Design.

Designed objects, like a car, are easily recognized because we could reasonably ask – what is the purpose of this object? – whereas that question makes no sense for a natural object, like a rock. The purpose of a car is to move people safely from A to B and it boasts important features – like a steering wheel and seat belts – that make it more likely that the car's purpose will be fulfilled. These features are certainly not accidents – they betray the existence of the unseen person who designed the object and thought hard about how it would work best. So, if we found a designed object on Mars, perhaps a weird-looking watch that could tell Martian time, we could be pretty sure that an intelligent alien designer was hiding somewhere.

At first glance, living things seem to share a great deal in common with designed objects. They too appear purposeful, although the purpose of a lion is simply to survive and produce more lions. But, like the car, a lion has a suite of features, called *adaptations*, that help it to achieve that goal – stocky legs, large pointy teeth and a killer instinct. So, must it too have a hidden designer?

While the bishops of the nineteenth century said yes, the theory proposed by Darwin and Wallace offered an alternative explanation. They suggested that a force called natural selection had moulded and shaped creatures over long periods of time, leaving them with the *appearance* of good design, despite the absence of a designer. Theirs was the first theory to offer a convincing natural explanation for the wondrous variety and brilliance of the organisms on our planet and it revolutionized the way we think about the natural world.

The theory of evolution by natural selection rests on three crucial assumptions. We need to examine these very carefully and decide whether or not we believe them to be true. If we are convinced, then these three assumptions also lead to an inevitable conclusion – that species will evolve to become better adapted to their environments. It's easier to see how it works with a concrete example, so let's use the peppered moth, as it's often seen as a classic example of natural selection in action.

1. Individuals have different features and many of these differences are heritable (passed from parent to offspring). In the case of the peppered moth, the key feature is wing colour. Some moths have speckled wings, while others have black wings, and we now know that these differences were caused by a mutation in the genome and so were heritable.

2. More offspring are born than the environment can support, so some will survive and thrive, while others will die before they can reproduce. In the case of the peppered moth, many are eaten by birds and so not all moths survive and lay eggs.

3. The heritable features of individuals affect the chances of dying or surviving and/or the number of offspring that they produce. During the Industrial Revolution, black moths were more likely to survive (and so have more offspring) than speckled ones because they were better camouflaged from birds against sooty tree trunks.

And now for the conclusion: *Species will evolve to become better adapted to their environments.* Over time, the number of black moths in the woodland will increase, while the number of speckled moths will decline, so on average, the peppered moth will be better adapted to its sooty home. And in a nutshell, this is natural selection in action.

Darwin called this process natural selection because some individuals were being selected over others. It reminded Darwin of *artificial selection*, where an animal breeder consciously chooses to breed only from animals with characteristics that they like. Of course, natural selection has no skin in the game – it blindly favours whichever features happen to work. For this reason, natural selection has led to some behaviours that humans find unpleasant, such as aggression or even cannibalism (the female praying mantis often eats her mate).

Darwin also used the term *fitness* to describe how natural selection might view individuals. For individuals to be favoured by natural selection, they need to survive *and* produce surviving offspring that will themselves be successful. The fitness of an individual is a way of rolling together all these different aspects of success, and it's sometimes approximated as the number of grandchildren that an individual can produce.

Both Darwin and Wallace individually spent many years refining their assumptions and collecting evidence to support them. Even so, their understanding was inevitably incomplete because they had no knowledge of genomes or mutations. The science of genetics and the discovery of the structure of DNA heavily bolstered the theory, so let's take a closer look at the evidence that supports each of the three crucial assumptions on which the theory rests.

1 Individuals have different features and many of these differences are heritable (passed from parent to offspring). Darwin and Wallace may have known nothing about genetics, but they knew enough about breeding domestic animals to know something about heritability. If a dog breeder wants to produce a litter of sausage dogs, it's advisable to breed from two sausage dogs, and not cross a great Dane with a poodle. Today, the science of genetics has revolutionized our understanding of inheritance. The genes that cause particular features have even been identified in some species (like coat colour in dogs), although there are many genes whose functions we still don't understand.

2. More offspring are born than the environment can support, so some will survive and thrive, while others will die before they can reproduce. The truth of this statement is easily seen – a single female codfish can lay up to 2.5 million eggs every year, while there are estimated to be fewer than one million adult codfish in the North Sea. Unfortunately, this means that many young animals will not survive to produce offspring of their own.

3. The heritable features of individuals affect the chances of dying or surviving and/or the number of offspring that they produce. The fitness of individuals has been examined in countless different organisms and it's clear that mutations matter. In the case of the peppered moth, experiments show that black moths survive better than speckled ones when placed on blackened tree trunks, and so have higher fitness in polluted woodlands.

In the end, natural selection is powerless without mutation. It may seem that the peppered moth was lucky that the black mutation arose at just the right time, but it's almost certain that the same mutation had arisen previously in unpolluted woodlands and the unlucky carrier was immediately eaten by a bird. And, while carefully drawn plans exist for designed objects, no plans will ever be found for the creatures on our planet. Unlike an animal breeder, natural selection does not set out with a goal in mind, nor does it 'try' to save species from extinction. It simply acts on whatever variation is present in a particular time and place, and no one can say which direction it might take or what creatures might emerge in the future.

Testing Darwin's theory doesn't just mean examining the assumptions on which it rests. It can also involve watching natural selection in action and observing the outcome – do we indeed see that species become better adapted to their environments, as the theory predicts? And experiments of this kind can be done either in nature or in the lab.

In 1988, Richard Lenski set up 12 glass bottles filled with all the things that cells need to grow and reproduce, and introduced a few genetically identical bacteria into each one. Populations of bacteria grow by cell division, and to stop them outgrowing their homes, just 1% of the bacteria were transferred into a new bottle with fresh food every day. The bacteria from the 12 different bottles were never allowed to mix and the daily transfers continued until the experiment was paused due to Covid in 2020.

Under the conditions in the bottles, each bacterial cell can copy its genome and turn itself into two cells in around 3.5 hours. Although this isn't particularly fast by bacterial standards (the conditions used in the experiment aren't ideal), it meant that by 2010, the bacteria in each bottle had made it through 50,000 generations, which would have taken more than one million years if Lenski had used humans instead of bacteria.

Over such a large number of generations, we might expect natural selection to act and ensure that the bacteria are 'better' at living in bottles, but how can this be tested? Normally, we have no way of knowing whether animals that died hundreds or thousands of years ago were any better or worse than their modern descendants. But bacteria are different. They can simply be frozen and resurrected when wanted, so we can compete the ancestors of the experiment against their descendants, and the results of these trials are quite amazing.

In all 12 populations, the descendants outcompete their ancestors every time, because their fitness has increased. After 20,000 generations, the cells from all twelve bottles were larger and grew 70% faster than their resurrected ancestors, and it didn't stop there. After 50,000 generations, cells were still able to outcompete ancestors retrieved from the same bottle 10,000 generations previously, so fitness was still increasing, although not as quickly as it had over the first 10,000 generations. But if all the bacteria were genetically identical at the start of the experiment, how is this possible?

Although the bacteria were genetically identical at the start of the experiment, mutation soon provided the necessary variation on which natural selection could act. Most of the mutations would have been harmful and so were eliminated, but a small number clearly allowed the bacteria to grow faster, and natural selection ensured that the bacteria carrying these mutations survived and thrived. All 12 populations have gone the same way, so it's no fluke – simply a logical consequence of the three assumptions laid out above – but there are a couple of interesting differences.

In just one of the 12 populations, a mutation arose that allowed the numbers of bacteria to massively increase. The mutation allowed the bacteria to feed on one of the ingredients in the bottle that their ancestors couldn't access, and this clearly gave them much higher fitness. But, despite its enormous advantage, this mutation simply has not appeared in the other 11 bottles, so it's worth remembering that

Lenski's Evolution Experiment with Bacteria

24ᵗʰ February 1988: 12 identical flasks are seeded with a small number of identical bacteria

Within each flask, the bacterial population grows

24 hours

Each day, a small number of bacteria are transferred to a new flask, where they keep growing, but populations are never allowed to mix

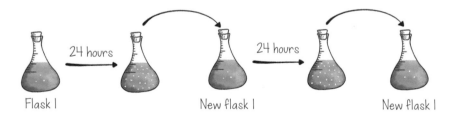

Flask 1   New flask 1   New flask 1

After 10,000 generations, the bacteria in each flask have evolved. The cells are larger and they divide more quickly

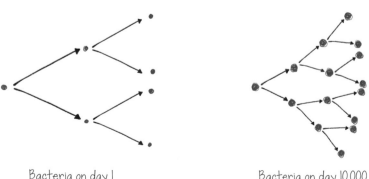

Bacteria on day 1   Bacteria on day 10,000

mutation is a chancy process and particular outcomes can never be guaranteed.

Although natural selection always acts to increase fitness, it can sometimes lead species in unexpected directions. As humans we have pretty strong feelings about what is 'better', but it's perfectly possible for species to lose features that we might think are highly desirable. And while a designer might be more careful to retain features that could be of use in the future, natural selection is relentlessly short-sighted, as the story of flightless birds reveals.

The desire to fly like a bird has captivated humans through the ages – from ancient storytellers to modern aeronautical engineers – and anyone who has gazed at a swift flashing past or an eagle soaring overhead has surely felt a twinge of envy for their almost supernatural aerial abilities. Only four groups of animals have ever successfully mastered powered flight – insects, pterosaurs, birds and bats – so, clearly flight is a rare gift. And yet, there are more than 60 species of flightless birds in the world from a total of more than 10,000. Most of them live on islands: there is a flightless cormorant on the Galapagos Islands, a flightless duck on the Falklands and a flightless woodhen on Lord Howe Island. But the real Mecca for flightless birds is New Zealand, with a whopping 16 species, including a parrot and, until recently, a tiny songbird. So why did these birds give up their finest feature?

Lying in the Indian Ocean, a small remote island called Aldabra is home to a unique flightless waterbird, the Aldabra rail. Alongside the rails, Aldabra hosts around 100,000 giant tortoises, but there are no mammals (with the exception of a few bats). Most of Aldabra's animals arrived there from Madagascar, which lies around 400 km away, and is filled with mammals, including lemurs and the cat-like predatory fossa. None of Madagascar's mammals have made it to Aldabra because mammals are not good at travelling long distances over the ocean. But, one of the many gorgeous birds to be found on Madagascar is the white-throated rail, and it looks very similar to the Aldabra rail, except that it can still fly.

The flightless Aldabra rails are descended from one or two Madagascar rails that one day flew to Aldabra. They might have taken a wrong turn or got lost in a storm, but once there, they quickly lost the power of flight. Indeed, we now know that this happened, not once, but twice, in just the last half a million years or so.

The average lifespan of a species is between 1 and 10 million years

So, has evolution 'gone wrong' or is there a good reason why birds on islands might lose the ability to fly?

Imagine a mutation that makes a rail a bit more reluctant to fly, perhaps by making its wings a little shorter. On Madagascar, the bearer of this mutation would rapidly become food for a fossa – but there are no predators on Aldabra. Indeed, on Aldabra the owner of such a mutation might gain an advantage over fellow rails. Flight demands fuel – a lot of it – and for birds, fuel means food. By expending less energy on flying, our mutant bird potentially has more reserves to see it through hard times, and it can perhaps grow larger and win crucial battles for food.

It's exactly this kind of short-term advantage that causes a mutation to sweep through a population and could swiftly lead the rails down a path towards a totally flightless state. Indeed, as many island birds all over the world have evolved to be flightless, we can be confident that the advantage is widespread. But, unfortunately, there is a dark side to the story of flightless birds that reveals how natural selection's relentless focus on short-term advantage can leave species high and dry.

In 1492, Vasco da Gama set sail from his native Portugal intent on finding a sea passage to India. His success opened up a regular trading route from Europe, around the tip of South Africa and across the Indian Ocean to the fabled Spice Islands (modern Indonesia). On these voyages, previously uninhabited islands – like Mauritius and Aldabra – were 'discovered' by Europeans and ruthlessly exploited.

Along with a desire for untold riches, the sailors plying these routes brought a surprising selection of passengers with them on their long and often hazardous journeys. Some, like goats and pigs, were invited guests, which were then released onto small islands to provide food for the return trip. But others, like rats and mice, were unwelcome stowaways that infested ships whether sailors wanted them or not. So, to keep the rodent problem under control, sailors also kept cats on board. And all these animals made themselves at home on islands where no mammal had ever previously set foot.

For flightless birds, the arrival of mammals was nothing short of disastrous. Being flightless might have been a great strategy when mammals were absent, but it left birds utterly unable to cope when they finally arrived: New Zealand alone lost 16 flightless birds, and the dodo of Mauritius didn't survive long once sailors turned up. Fortunately, the Aldabra rail had a lucky escape. Aldabra consists of several small islands separated by deep-water channels, and on one of these, cats never arrived. The rail survived there and has since been reintroduced to other islands, following the removal of cats.

It's reasonable to ask why species don't simply re-evolve lost features, but – while not impossible – this is much harder than losing them in the first place. Mutations occur by chance, and there are lots of ways to break a protein – all sorts of mutations will convert one amino acid into another and destroy the protein's proper shape. But restoring a lost feature like flight would require reversing each specific mutation that had led a bird down the path to flightlessness, and this is fantastically improbable. The unhappy result for many flightless birds, including the dodo of Mauritius, was extinction within decades of human arrival. In an age of unprecedented environmental change, it's worth remembering that natural selection can't always save the day.

The evolution of powered flight was undoubtedly a momentous event in the story of birds, but some innovations have had dramatic consequences for more than one life form. For the first approximately 3.5 billion years, cells on our planet lived alone, but eventually they began to gang up, giving rise to complex multicellular beings like plants, fungi and animals. But building a tree, a mushroom or a lion requires enormous numbers of cells to co-operate seamlessly – and persuading them to do so is a much more serious challenge than getting our feathered friends off the ground.

The key feature of multicellular organisms is that cells within the same body co-operate and work together. If they didn't, then the body they share is very unlikely to be successful. Indeed, so closely do cells work together, that many of them are even prepared to sacrifice themselves for the good of the body they inhabit. But if natural selection concerns itself only with fitness, then why should some cells give up their fitness in order to help others?

To unravel this problem, let's begin by looking at a situation where this kind of co-operation has NOT evolved. No doubt we've all watched wildlife films in which lions attack a herd of wildebeest. Often there are thousands of wildebeest and only a small number of lions, and we've probably all asked ourselves why those thousands of wildebeest don't just gang up and kick the lions to death – because although a few might get hurt – the survivors could then live in peace forever.

To see why this doesn't work, consider a single wildebeest called Galahad who is carrying a novel mutation. The mutation changes Galahad's behaviour, causing him to charge in and help any other wildebeest who is being attacked by a lion, regardless of the risks to himself. So, will the Galahad mutation spread?

The problem for Galahad is that he is very likely to injure himself and die at a much younger age than his friends that lack the Galahad mutation and who simply run away whenever the lions attack. And when Galahad dies without offspring (no doubt mourned by his many friends and admirers), so perishes the mutation that caused his noble behaviour. The upshot is that wildebeest continue to prioritize their own individual survival and leave others to die at the hands (or paws) of lions – no matter how upsetting this may be for viewers of nature documentaries.

But there is an important difference between a herd of wildebeest and the cells within a single organism. Whether plant, fungus or animal, every complex multicellular organism starts life as a single founding cell that divides again and again, so all the cells within the body are *clones* that share an identical genome – and this is absolutely crucial to their evolution.

Imagine a simple multicellular creature filled with co-operating cells that can all still reproduce and give rise to a new creature. But now, a mutation occurs in the founding cell of one of these bodies, which is passed on to all of its descendants as the body develops. Although *all* cells carry the mutation, it only causes *some* cells to forego reproduction, perhaps because they specialize in doing something else – like stinging predators to death – that causes them to die in the process. These cells are a bit like Galahad – they perform heroic acts to support the other cells, but at the cost of giving up their own fitness. So, can this mutation spread?

The answer is yes, as long as the mutation makes the whole body more successful than other bodies that lack these specialized 'Galahad' cells. Although the offspring are now produced only by the reproductive cells, *they will all carry the mutation*, because the reproductive cells have exactly the same genome as the stinging cells.

The Galahad Mutation in a Multicellular Being

A founding cell

It divides to form many identical daughter cells that stay together and form a simple multicellular body

After a short time feeding, the body falls apart and all the daughter cells can found a new multicellular body

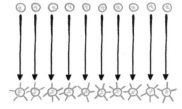

A founding cell that contains a Galahad mutation

It divides to form a simple multicellular body BUT some cells, which we can call Galahad cells, become stinging cells, which defend the body against predators and die

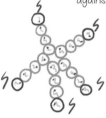

When this body falls apart, ONLY the non-Galahad cells can found a new body. BUT these non-Galahad cells all carry the Galahad mutation

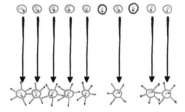

In a world filled with predators, the Galahad mutation can spread

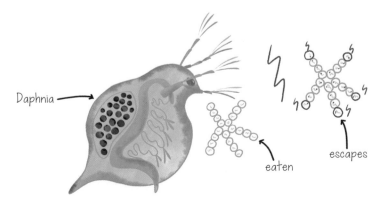

Daphnia

eaten

escapes

So – unlike poor thankless Galahad – this mutation *can* be successful, and multicellular beings filled with self-sacrificing cells can (and do) evolve, and we will see several examples in future chapters.

All complex multicellular organisms start their lives as a single cell, which suggests that sharing the same genome is an essential feature of close co-operation. But we also see co-operation in other situations – just think about the effort that parents expend on their children. It's easy to think that it's perfectly 'natural' for a wildebeest mother to protect her own calf rather than one belonging to some other wildebeest, but this too requires some explanation.

All organisms carry a genome inherited from their parents – so there is a good chance that the offspring will inherit any mutation that the parent has acquired. If a mutation arises that causes a mother to charge in and protect her own baby, then it might spread through the population if her actions make it more likely that her offspring survives – and with it the mutation that she passed on. Of course, mothers have to be careful not to run too great a risk. If a mother dies in the process of defending one offspring, then her reproductive life is over, so she can't be too reckless. The upshot is that many animals are quite good at assessing the relative risk to themselves and to their offspring. Female wildebeest will defend their calves more vigorously against a cheetah – which is no real threat to the mother – than against a lion, that can easily take her down.

In the natural world, *altruism* – which means to sacrifice yourself for the good of others – has pretty sharply defined limits, and these limits are generally defined by relatedness. Honeybee workers gather pollen to raise their hive-bound sisters, while many different bird species like scrub jays or long-tailed tits help their parents to rear more chicks. But when these cases are closely examined, the helpers and those being helped are nearly always related to each other, so there's a good chance that they share the mutation that causes the helping behaviour. And by helping a relative, they are unconsciously helping that mutation to spread.

It's somewhat disappointing, but natural selection is much less likely to favour individuals (like Galahad) who perform altruistic acts to perfect strangers. But, that doesn't mean that it never happens. Good examples of this kind of behaviour are found in animals, like meerkats, who live in co-operative groups. Complete strangers will sometimes join a group and help to rear the offspring of the dominant pair. This seems like a Galahad effect, but the unrelated animals are waiting for a chance to claim dominance and have offspring of their own, and their alternative – living alone – is almost certainly doomed to failure. So, animals might help strangers, if by doing so they can increase their future fitness.

The unhappy dodo and the mythical Galahad tell us that natural selection does not act for the good of species, it can only pursue individual advantage. Humans sometimes struggle to accept this reality and find the selfish behaviour of animals hard to understand, but it doesn't mean that we are forced to emulate them. We are conscious beings and can choose a different path for ourselves, no matter what natural selection might think about it. Indeed, it would be a very strange society that only valued people according to the number of children that they successfully raised.

Over the course of the Earth's history, enormous numbers of species have come and gone, as some evolved beyond recognition, while others failed to adapt and disappeared forever. Nothing has given us greater insight into these lost worlds than *fossils* – the remains of plants and animals, or traces of their presence – that have turned to stone. But although humans have been finding fossils for thousands of years, their significance has only been appreciated much more recently.

One of the great figures of palaeontology is Mary Anning, who lived around the same time as Darwin and Wallace in Dorset, England. She worked the cliffs near her home collecting fossil specimens to sell, although many of her customers had strange ideas about what they were buying. Her best-selling fossils were the beautiful coiled shells of *ammonites* – traditionally thought to be the remains of a plague of snakes, turned to stone by St Hilda of Whitby. But during Mary's lifetime, supernatural explanations were gradually being abandoned as a scientific revolution swept through Britain.

At the turn of the nineteenth century, even die-hard creationists began to accept that there was evidence of *extinct* creatures – ones that could no longer be found anywhere on Earth. One of the first extinct animals to be properly described was the woolly mammoth, a relative of modern elephants that can be found beautifully preserved in Arctic tundra. The mammoth was followed by the discovery of other extinct mammals, like the giant South American ground sloth, so it seemed that there must have been at least one former world with rather different creatures. But at least they were similar to the ones around us now.

As Mary worked the cliffs of Dorset, she revealed evidence for an even older time in which mammals barely featured. Mary found the first *ichthyosaurs* and complete skeletons of long-necked *plesiosaurs* – large marine reptiles that look nothing like any animal alive today. She also excavated one of the first *pterosaurs* – winged reptiles that flew using a thin membrane supported by an enormously long, bony finger. Mary's findings caused an absolute sensation – could there really have been a world populated by reptilian giants?

As it became clear that extraordinary monsters had once roamed, roared and battled it out on our planet, people couldn't get enough of it. Bitter rivalries broke out among scientists as they haggled over new fossil finds and the right to be associated with them. In Europe, excellent fossils of large marine reptiles, like those found by Mary, were some of the first to emerge, but these were soon eclipsed in the public imagination by sensational discoveries of familiar *dinosaurs*, like *Tyrannosaurus* and *Diplodocus*, in North America.

Despite the excitement, dinosaurs were still viewed through the lens of Victorian ideas of progress. Even today, the word dinosaur is used to label something as hopelessly out of step with the times, with the implication that the real dinosaurs deserved their fate. At the time of their discovery, newly unearthed skeletons were bolted together in unconvincing postures, where they looked, frankly, clumsy and inept. The message was clear – these outdated tail-draggers were a failed experiment in evolution and wouldn't stand a chance against a sleek modern human.

But since their discovery, our understanding of evolution, and of dinosaurs, has radically changed. Dinosaurs were honed by natural selection over millions of years and were superbly adapted to the environments of their day. And other evidence – besides their bones – has been used to reconstruct their lives and behaviours.

The pterosaur, *Quetzalcoatlus*, had a wingspan of 10 to 12 metres

Fossilized poo has revealed that *Tyrannosaurus rex* was a bone-crunching predator of terrifying proportions that often preyed on juvenile dinosaurs; analysis of tracks has revealed that *sauropods* like *Diplodocus*, far from dragging their tails, probably cracked them like bull-whips to intimidate rivals; and discoveries of multiple nests at the same site reveal that some dinosaurs, like *Maiasaura*, nested in large colonies, rather like modern seabirds, and cared for their young. One thing's for sure – although a good athlete might comfortably outrun a dinosaur – most modern humans wouldn't last long in the real Jurassic Park.

Dinosaurs may fire our imaginations more than any other type of extinct animal, but the fossil record can tell us much more about the general pattern of evolution and the history of life on Earth than the last supper of a doomed *Tyrannosaurus rex*. Despite being incomplete, careful examination of the fossil record allows us to retrace evolutionary journeys and see when and how key innovations emerged. And it's also clear that our Earth is far more ancient than anyone in the nineteenth century could possibly have believed.

Accepting Darwin's theory means getting to grips with the enormity of the changes wrought by natural selection. All life can be traced back to a common ancestor, which means that each creature must have evolved from ones that went before. Extinct plants and animals look quite different from those around us today, and while most of us aren't troubled by the idea that a moth can change colour over a few decades, the idea that a fish could turn into a frog is quite another matter. But there is only one crucial difference between these two events – a precious commodity that we all want more of: time.

Darwin knew that if all life was descended from a single-celled ancestor, the Earth must be immensely old. Accurate dating would have to wait until the twentieth century, but during Darwin's lifetime geologists pushed back the age of the Earth from thousands to millions of years, and we now know it's roughly *4.5 billion* years old – a span of time that is extremely difficult for any puny human to comprehend. Various comparisons have been made – if we reduce the Earth's history to 24 hours, then humans only appeared at the last stroke of midnight – or if we reduce it to one year, then humans appeared at around 11.35 pm on 31 December. But, none of these comparisons allow us to imagine what millions of years really feel like.

Life emerged very quickly once the early Earth cooled, although few direct traces can be found in rocks so old. This life was all single-celled; indeed, if we compress life on Earth into one year, the first proper animals don't appear until around 18 November, so single-celled life has dominated the planet for most of its history. The appearance of the first animals 541 million years ago is a sudden event (geologically speaking) and so is dubbed the *Cambrian explosion*. It marks the first recorded appearance of animals that could see, swim and hunt. Crucially, these animals had visible means of support, like bones, shells, skulls and teeth, so from this point forward, the fossil record – although incomplete – is richly informative.

The Cambrian marks the start of the eon of visible life, and we divide it into three great eras – the *Palaeozoic* (ancient life), the *Mesozoic* (middle life) and the *Cenozoic* (recent life). These three eras differ in length. If we return to our one-year timeline, the Palaeozoic began on 18 November and ended on 12 December when the Mesozoic took over and ran for a full two weeks before ending on Boxing Day (26 December) and handing over the last five days of the year to the upstart Cenozoic.

The Palaeozoic oversaw the greatest changes to life on Earth, partly because it's the longest, and partly because it began when there was practically no life on land at all. Over the course of more than 250 million years, the first plants emerged on the barren continents, although they were soon followed by a range of animals, including insects and animals with backbones (vertebrates). Indeed, so successful were these early colonizers, that by the end of the Palaeozoic, there were great forests standing 30 metres tall, while fantastic reptiles prowled below.

The Mesozoic came next, and like all eras is further divided into geological *periods*, which are shorter, although each can still last for tens of millions of years. Some geological periods are practically household names, like the Jurassic – the second of three periods that make up the Mesozoic. Most of us correctly associate the Jurassic with dinosaurs, although Mary's ichthyosaurs roamed the seas and pterosaurs filled the skies (and these are not dinosaurs).

The Cenozoic is the briefest era and was marked by the dominance of mammals and birds. At the end of the Cenozoic, the Earth was hit by multiple glaciations or ice ages, which caused many mammals to increase in size. The last ice age only ended around 12,000 years ago, which is why we can still find the perfectly preserved remains of mammoths in the Siberian tundra. Although many ice-age giants have now vanished, there are still a few survivors in the modern world, including the inappropriately named musk ox (which is not a cow but an incredibly bad-tempered sheep).

Some of life's evolutionary journeys can be traced through the eras by following sequences of fossils. One exciting example is the evolution of four-legged vertebrates (called tetrapods) from fish during the Palaeozoic. A wonderful series of fossils has been found, beginning with an air-breathing fish with muscular fins, followed by an animal with a fish-like rear end but leg-like forelimbs, and culminating in four-legged semi-aquatic animals. But as well as seeing gradual changes in the fossil record, we can also see catastrophic events and we use these to mark the boundaries between intervals of geological time.

The three great eras within the eon of visible life are separated by two devastating *mass extinctions*, in which whole suites of plants and animals disappeared forever in a geological instant. The mass extinction that brought the Palaeozoic to a close around 250 million years ago was so dramatic that it's often called the Great Dying. It was caused by massive sustained volcanic activity that raised global temperatures over a short period of time, and although it was the most severe mass extinction in Earth's history, it's the mass extinction that wiped out most of the dinosaurs at the end of the Mesozoic that tends to grab all the headlines.

This mass extinction had an extra-terrestrial cause – a massive asteroid. The impact crater where it crashed into the Earth lies in the Gulf of Mexico, at a site called Chicxulub (pronounced: *Chicks-uh-lub*), and at around 100 kilometres across, we estimate that it was made by an asteroid somewhere between 10 and 80 kilometres wide that struck with the force of more than one billion atomic bombs (of the kind dropped on Hiroshima). Excavations have revealed that granite from 20 kilometres down was forced above ground and the surface melted over a wide area, obliterating life across the entire North American continent and causing wildfires to rage around

Geological Time

HADEAN

First cells

First photosynthesis

Great Oxidation Event

4

4.5 bya

3

Cambrian explosion

Anomalocaris

Coal-forming forests

300

500 mya

PALAEOZOIC

Lystrosaurus

400

Rhynie cherts

Early tetrapods

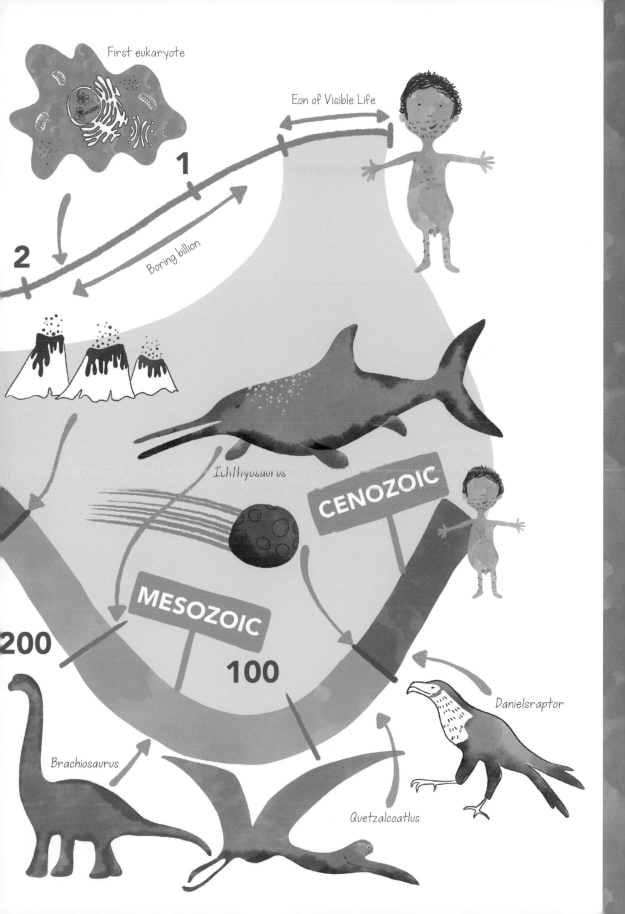

First eukaryote

Eon of Visible Life

1

2

Boring billion

Ichthyosaurus

CENOZOIC

MESOZOIC

200

100

Brachiosaurus

Quetzalcoatlus

Danielsraptor

the globe – perhaps the closest the Earth has come so far to Armageddon.

Following a mass extinction, the ravaged Earth may take millions of years to fully recover. Sometimes entire groups are swept away, like the ammonites that perished at the end of the Mesozoic, but this leaves opportunities for others. Natural selection quickly gets to work on the survivors, sculpting them to fill the gaps left by those who have been written out of life's play, and ensuring that they will thrive in the brave new world in which they now find themselves.

Mass extinctions vastly reduce the number of species on Earth. But intriguingly, the fossil record appears to show that, despite these setbacks, there are more species on Earth today than at any other time in the past. So, if today's world is bursting with an unparalleled bounty of life, where did they all come from?

Understanding the origins of new species requires us to look closely at how multicellular creatures, like birds and bees, reproduce. We know that each one starts life as a single cell, which then divides again and again, but because all cells are products of cell division, that first cell must have come from somewhere. Most of us are probably aware that the first cell of our own bodies came from our parents. But why are two parents required to kick-start each precious offspring?

Chapter 3
SEX
Running to stand still

In a popular song of the 1920s, the legendary songwriter Cole Porter urged people to fall in love, claiming that we should follow the example of birds, bees and educated fleas. Whether or not non-human animals are capable of complex emotions like love is definitely contentious. Some birds are certainly more faithful to their partners than many humans, but after mating, neither bees nor fleas show any further interest in their would-be husbands or wives (regardless of educational status). In fact, the behaviour shared by all these animals – and indeed nearly all multicellular creatures – isn't love, but sex. And while it might be the normal mode of reproduction for fleas and humans alike, many aspects of sex continue to puzzle biologists.

Consider a single-celled organism, like a bacterium. Normally, such cells reproduce by dividing themselves into two identical daughter cells that go on to lead independent lives. But while this works perfectly well for bacteria, multicellular creatures can't just carve themselves in two.

As we have seen, larger, more complex creatures evolved from unicellular ancestors. But the enormous teams of cells that animate these bodies are always descended from a single founder, so they all share the same genome. This allows close co-operation, and in most bodies, a few cells are set aside to produce the next generation, while the rest are doomed to die with the body they inhabit.

Most multicellular creatures don't begin to reproduce until they reach adulthood, at which point the reproductive parts spring into action and start to produce sex cells or *gametes*. In both animals and plants, these come in two different types: *egg cells* that are generally immobile, and much smaller *sperm cells* that can actively swim (although in most plants, the sperm cells can't swim and are encased in tough pollen grains, which are only released when the pollen arrives in the right place on another plant). But there is something highly unusual about the way all sex cells are generated.

To produce a new cell of any type, an existing cell must divide, and this normally yields two daughter cells that each receive a complete copy of the genome. But the genomes possessed by all familiar animals – fish, birds and humans alike – share a strange secret. The genome is a book of instructions, but in multicellular creatures it's actually a double-act – two copies of essentially the same instruction manual that work together so seamlessly that most of the time we can ignore its double nature.

Like all other cells, egg and sperm cells are a product of cell division, but *un*like other cells they only receive a single copy of the manual, which leaves them short-changed and unable to function. Unless of course they can team up with another sex cell and pool their genetic resources.

Rounding the bend of a sparkling stream, somewhere in the highlands of Pakistan, a group of zebrafish have heard the call of Cole Porter. Fighting hard against the current, several small males are doggedly pursuing a group of large, plump females, and as they enter the shallows, the males finally catch them up. For a brief moment, the adult fish engage in a bout of vigorous courtship as the males shimmy up alongside the females. This frenzied wriggling stimulates the females to release a burst of eggs into the water and the males respond by releasing a cloud of sperm. But while the adults are now free, the cells they leave behind have unfinished business.

A zebrafish egg consists of a single enormous cell perched atop a nourishing yolk. Yet despite its impressive size, the egg isn't quite the cell it should be. Carrying only a single copy of the zebrafish manual, it desperately needs to find a partner, so the egg cell must wait, frozen in time, until the arrival of Prince Charming.

Prince Charming is the rather less impressive zebrafish sperm cell. Produced inside the bodies of males, sperm cells are much smaller than egg cells and carry only minimal kit – but they are equipped with a long tail for frantic swimming. Crucially, each sperm cell also clutches a single copy of the zebrafish manual, but delivering it will not be easy. Because of their tiny size, millions of sperm cells were produced by the males while the enormous egg cells are few and far between. Just as the male fish had to scramble to keep up with the females, so the sperm cells now take part in a desperate race of their own. Each egg cell will only allow a single sperm cell to enter and there are no prizes for coming second.

On reaching the egg, the winning sperm cell punches a hole in the egg cell membrane and pushes its half-genome inside. Called *fertilization*, the joining of the egg and sperm cells allows the two half-genomes to combine and form a fully functioning founding cell, which can begin to divide and form a brand-new fish.

Strange though it seems, it's hard to imagine a more significant event in the life of any organism. We may endlessly debate whether egg preceded chicken, but this is indisputably the first step on the road that leads from one to the other.

So, let's recap the essentials of *sexual reproduction*. First, individuals with perfectly good genomes produce sex cells which only contain half the information they need. Next, they are let loose so that sex cells from different parents can combine their half-genomes and end up with a whole genome again. Finally, the new founding cells can begin to undergo normal cell division and produce brand-new individuals. This whole fiasco – splitting up two perfectly good genomes and then sticking them back together again in new and different combinations – is really what Cole Porter was singing about. And you could certainly be forgiven for wondering what the point of it is.

All animals produce both egg and sperm cells, but not all animals have separate male and female sexes. Many animals, like the common snail found in European gardens, are *hermaphrodite* and can produce both male *and* female sex cells from the same body. This arrangement doesn't mean that snails don't seek out mates – indeed, they are passionate creatures – it simply means that each snail will play both male and female roles in the courtship and mating that follows.

When two garden snails encounter each other, they use chemical cues to decide whether or not their potential partner is really worth the effort. If they decide to go ahead, then an elaborate ritual ensues that can last for several hours. The two snails intertwine their bodies, aided by the production of copious amounts of slime, and use their tentacles to investigate each other's faces. If all goes well, they will transfer packets of sperm into each other's bodies that will be stored and used to fertilize their own eggs.

In an extraordinary twist, many common snails also fire love darts – a sharp, hardened miniature arrow – into the bodies of their partners. The dart lodges within the partner snail and secretes chemicals called *hormones* that make it more likely that the transferred sperm will actually be used – rather than the first snail

Sexual Reproduction

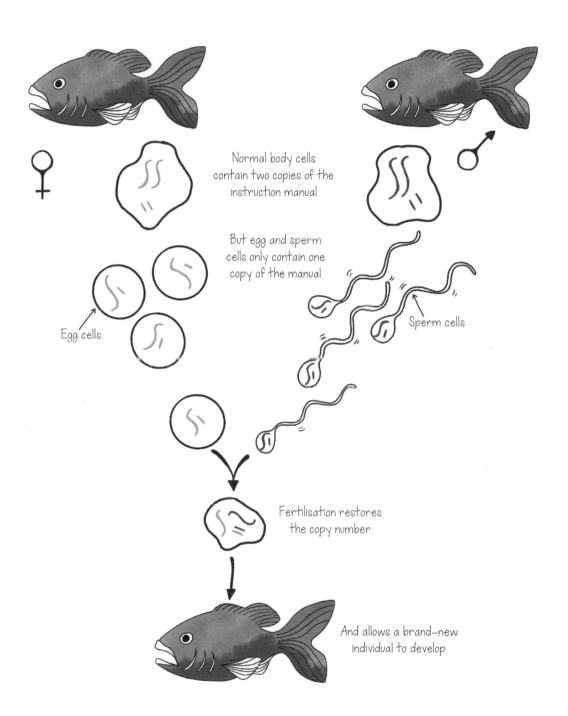

Normal body cells contain two copies of the instruction manual

But egg and sperm cells only contain one copy of the manual

Egg cells

Sperm cells

Fertilisation restores the copy number

And allows a brand-new individual to develop

simply falling back on self-fertilization. Self-fertilization is always possible for hermaphrodite species, but the partner snail's fitness is enhanced by fathering another snail's offspring, so it wants to do everything in its power to prevent self-fertilization from happening – hence the love darts. But for most vertebrates, like fish, lizards, birds and mammals, self-fertilization is simply not an option.

Most vertebrates have separate sexes: females to produce egg cells and males to produce sperm cells. The need to secure a sexual partner means that significant chunks of the lives of many animals are dedicated to attracting a mate, and much of the behaviour that fills our TV screens is centred on animals doing just that. But the problem for biologists is why animals (and other multicellular creatures) expend so much energy obtaining a half-genome from some other individual, when all individuals possess a perfectly good whole genome of their own.

Let's return to the zebrafish and explore for a moment what an *asexual* world (one without sex) might look like. In this sexless world, a female fish produces egg cells in the same way that she produces any other type of cell, so each one contains a copy of her entire genome. Because each egg cell already has all the required instructions, it can immediately start to develop into a new fish, without the need for a sperm cell to fertilize it. And because the eggs are identical to their mother, they all develop into females, so this population would be entirely free of males. So, if a mutation arose that converted a sexual female into an asexual one, do we expect her to be successful?

To answer this question, we have to compare her fitness to that of the sexual females with whom she is now competing. Each offspring produced by a sexual female only contains half of her genome, while the other half is contributed by an unrelated male. In fitness terms, this seems disastrous, as natural selection always favours individuals who maximize the transfer of their own genetic material to the next generation. Worse, half of her offspring are males, who can't lay eggs of their own. The upshot of these differences is that we expect the number of asexual females to increase twice as fast as the number of sexual females, and so easily win out, a phenomenon known as the *cost of sex*.

Amazingly, this sexless world is not science fiction. There are around 12 species of lizard and one (highly successful) snake that consist entirely of females who

The Cost of Sex

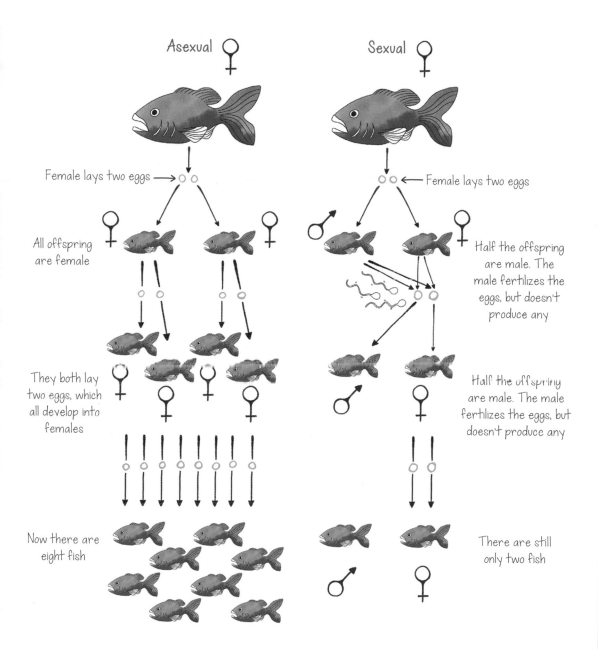

Note: Real fish lay more than two eggs, but the principle is the same: the population of asexuals can increase faster

reproduce asexually. But given that these are the exception rather than the rule, there must be a hidden downside to the sexless strategy. A clue to this disadvantage is revealed by noting that nearly all asexual species in the world today have evolved only recently (with very few exceptions within the last one million years). This suggests that asexual species struggle to persist over long periods of time, which in turn suggests that sex gives species some crucial long-term advantage.

The key difference between the sexual and the asexual strategies lies in the offspring. The ones produced by our imaginary sexless fish are all *clones* – genetically identical to each other and to their mother, because they developed from egg cells that were each handed a complete copy of their mother's genome. In contrast, the offspring of the sexual females, like brothers and sisters in the same human family, will all be different from each other and to their mother, and these differences are due to two important processes: recombination and fertilization.

Recombination takes place when egg and sperm cells are made using a specialised form of cell division called *meiosis*. The genomes of animals and plants are enormous, and so they are divided up and carried by multiple *chromosomes*. The number required varies greatly among species – a single copy of the human genome is carried by twenty-three different chromosomes, but some species have many more. As sexual species, we all carry two slightly different versions of the manual – one inherited from each parent – so the actual number of chromosomes in each 'normal' body cell is twice the single-copy number. This means that a human heart, liver or kidney cell contains forty-six chromosomes (2×23).

To form sperm or egg cells, the two copies of the manual must be separated. During meiosis, each chromosome lines up next to its opposite number, so in humans, we would see twenty-three pairs of chromosomes neatly lined up side by side. Within each pair, one of the chromosomes was inherited from the mother and the other from the father. Intimate contact between the two chromosomes in each pair allows recombination to take place: sections are exchanged to make brand-new versions of the manual – it's a bit like cutting and pasting between two very similar documents. Once finished, the pairs are pulled apart, leaving each human egg or sperm cell with twenty-three chromosomes, and a brand-new version of the manual.

Recombination

Sexual species possess two copies of the instruction manual

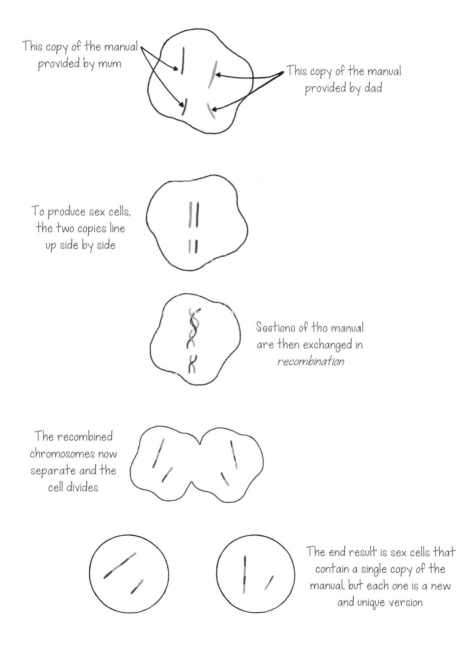

This copy of the manual provided by mum

This copy of the manual provided by dad

To produce sex cells, the two copies line up side by side

Sections of the manual are then exchanged in *recombination*

The recombined chromosomes now separate and the cell divides

The end result is sex cells that contain a single copy of the manual, but each one is a new and unique version

The second way that variation is introduced into offspring occurs during fertilization, when the sexual female's egg cells combine with a sperm cell produced by a male. Even if all the eggs produced by a single female are fertilized by the same male, recombination ensures that no two of his sperm cells will be exactly the same; and because his genome will be different from hers, the offspring of the sexual female are genetically much more diverse. But how can producing genetically diverse offspring offset the cost of sex?

Genetic diversity is important because the environment never stands still. We have already seen how the tree trunks in Northern England quickly blackened during the Industrial Revolution, allowing the mutant moth to prosper, and the Earth's climate has changed dramatically over millions of years (although it's rarely changed as rapidly as it is doing now). While the asexual female has to hope that the environment continues to suit offspring that are exactly like her, the sexual female is hedging her bets – *her* offspring are all different, so there is a better chance that at least one of them will be equipped to meet any new challenge that the environment might throw at them.

Still, natural selection is only really interested in short-term advantage, so we might rightly wonder whether the pace of environmental change is rapid enough to offset the cost of sex. But organisms aren't just battling the weather or their surroundings. One key aspect of the ever-changing environment is *other species*, and they are also subject to natural selection, which continually drives them forward. Whether as predator, prey or competitor, all species interact with others, and if the predator evolves new tactics – like the ability to run faster – then the prey needs to respond with new strategies of their own. And in one particular kind of interaction – those with *parasites* – sex seems to be crucial in allowing species to respond effectively.

Amazingly, close to half of all the species on our planet are parasites – happily exploiting the remaining, free-living half. Parasites live on or inside another organism and tend to be much smaller than the species they exploit. Unlike a predator that quickly kills and eats its victim, parasites spend hours, days, weeks or even years living in intimate association with their host, weakening it by consuming its resources. Most parasites don't kill their hosts, preferring to keep them alive for as long as possible, but some, like the larvae of tiny parasitoid wasps that consume the greenfly in the garden, are more merciless and usually destroy the host – leaving them somewhere between parasite and predator. So, before considering how sex prevents parasites from getting the upper hand, let's take a look at the variety of parasites on offer.

Parasites come in many shapes and sizes – all of them unpleasant and some of them, frankly, stomach-churning. Beginning with the smallest, viruses attack nearly all familiar species, and single-celled organisms, like bacteria, are well-known for causing disease in both animals and plants. Fungi also parasitize. Many plants are attacked by fungal pathogens, with evocative names like rusts, wilts, scabs and blights, which have the potential to devastate crop plants. At the other end of the scale, some parasites are enormous: a tapeworm that lives in the bowels of Arctic whales can grow to over 30 metres long and probably live for at least 20 years. Indeed, worms have been enthusiastic adopters of the parasitic lifestyle: flukes, hookworms, pinworms, whipworms, roundworms and tapeworms all make themselves at home in the blood or guts of various vertebrates, including humans. And worms aren't the only players in the parasite game.

Crustaceans are a group of animals that include crabs, lobsters and shrimps. Most crustaceans are free-living, but some use their claws to more sinister effect. The tongue-eating louse sneaks inside the body of a fish through its gills and attaches itself to the tongue, where it severs the blood supply, causing the tongue to fall off. The parasite then attaches itself to the stub of the tongue where it feeds on the blood or mucus of the unfortunate host. Incredibly, the fish continues to feed normally, apparently unaware that its tongue has been replaced by an evil shrimp.

Once safely inside the body of a host, parasites shed thousands of eggs or larvae, but these need to find a new body to inhabit. Direct travel from one host to the next is often impossible, so parasites have been forced to rope in unwilling accomplices. These take the form of additional hosts belonging to entirely different species that can help the parasite get back into the first host again.

Thanks to the global demand for sushi (raw fish), adults of the fish tapeworm are estimated to live in the guts of around 20 million humans, where they can grow up

250 -million- year-old shark poo contains the oldest known tapeworm eggs

to ten metres long and produce one million eggs every day. But the eggs have to find their way from one human to another, and this process begins when eggs are washed into nearby ponds and streams, where they are eaten by tiny shrimp-like crustaceans that mistake them for food. Inevitably, many of those crustaceans are themselves eaten by small fish and the small fish are snapped up by bigger ones – and incredibly, the juvenile tapeworm survives each of these transfers. Eventually, an unsuspecting human enjoying a day out fishing will catch one of these bigger fish, like a salmon or a perch, and if they don't cook it thoroughly, the juvenile tapeworm will have found itself a comfortable new home in which to spend its adult life.

To increase the chances of making a successful move from one host to another, some parasites engage in mind control. One parasitic fluke spends its adult life inside the guts of birds, shedding eggs into the bird's poo. But other birds don't eat bird poo, so the fluke needs an accomplice to help transfer its offspring into a new host – enter, the unsuspecting snail. Snails will happily eat bird poo, and with it the fluke's eggs. The larvae hatch out inside the snail and make their way into its eye stalk, where they bulge and glow, giving a fair imitation of a wriggling worm – irresistible to a passing bird. To make sure that birds don't miss out on this tempting treat, the larvae hijack the snail's behaviour, forcing it to sit out in the open, when it would much rather hide away in a cool damp spot (and perhaps indulge in a little courtship).

It's reasonable to ask how parasites get away with such blatant exploitation: after all, the host doesn't want them and will try to root them out. Given that parasites live deep inside the bodies of other animals and plants, we might expect the host to be in control; but parasites have evolved an extraordinary array of strategies to deceive the host and fight back against any efforts to remove them.

Most vertebrates have an *immune system* with specialist cells, whose job it is to recognize invading cells and flag them for destruction. The system must be fine-tuned or the animal might start to destroy itself, so all cells place signs on their membranes, called *antigens,* that reveal their identity. Cells that belong to the same body display antigens that mark them as 'self' and are ignored by the immune cells, which focus on actively hunting out 'non-self' potentially dangerous cells that display the wrong sign. Some parasites confuse the immune system by continually changing the signs that they display; others try to sneak under the radar, masking their own signs and pretending that they belong to the host; and a few go on the attack, sending false signals and stopping the development of specialist immune cells that are programmed to hunt down and kill invaders.

This intimate relationship between host and parasite is a result of *co-evolution*, where each species has evolved in response to the other. Natural selection will favour any change in the host that gives it an advantage over the parasite, and equally the parasite is undergoing natural selection to gain an advantage over the host. Attack and counter-attack can lead to an escalating arms race, in which host and pathogen attempt to outdo each other, leading to more virulent pathogens and better resistance in the host. But sometimes these interactions just go round in circles.

If a mutation arises that makes the host resistant to the parasite, then this will sweep through the host population. But with the hosts now resistant, the parasite is in danger of extinction. Fortunately for the parasite, mutations continually arise within its genome, and sooner or later, a parasite will emerge that can attack the resistant host. This new version of the parasite will undoubtedly be successful, but as it becomes more common, the host is under increasing pressure to respond.

It's entirely possible that a second mutation might arise in the host that reverses the effect of the first. This mutation could now spread, because the *original* host might be resistant to the *new* version of the parasite. But now a mutation in the parasite to restore *its* original genome will be favoured and, if it spreads, we will be back where we started – until a mutation arises in the host to make it resistant again. We think that these cycles can continue indefinitely, and now we finally see a clear advantage of sex. Under these conditions the sexual females leave their hypothetical sexless cousins on the starting blocks, as by shuffling their genomes through sex and recombination, the sexual population can track the changes in the parasite, and so keep one step ahead, whereas the parasite can easily overrun the asexual population.

The idea that sex evolved to prevent hosts being overrun by their parasites is often referred to as the *Red Queen* hypothesis. The name comes from Lewis Carroll's book *Through the Looking Glass*, where Alice (of Wonderland fame) is involved in a race. When Alice complains that they appear to be getting nowhere, the Red

Queen famously replies that in Wonderland it takes all the running they can manage just to stay in one place, while actually getting somewhere would require the impossible feat of going twice as fast. This image – running simply to stand still – perfectly describes the relationship between hosts and their parasites. They are locked in an endless battle of one-upmanship that can only end if one of them is finally driven to extinction by the other.

The Red Queen is just one possible explanation of why sex is so ubiquitous among multicellular creatures, and this question will no doubt continue to keep biologists busy. But regardless of whether or not sex evolved to keep parasites on the back foot, it has certainly had far-reaching consequences for the inhabitants of our planet. Our world would be a duller and less interesting place without sexual reproduction, because much of the colour and vibrancy we admire is a result of individuals trying to attract a mate. Unfortunately, there is also a downside to sexual reproduction, as much of the violence in the natural world can also be laid at its door.

Let's begin by slaying a few myths. Whatever the claims of national newspapers, males do not typically have more sex than females, nor do they have more offspring. Indeed, this would be impossible, because the double genome that we all carry requires precisely two parents – one of each sex. But, despite both sexes being equally successful on average, males and females that belong to the same species don't always look – or behave – in the same way.

In some species, males are either more glamorous or more violent than females and come suitably equipped with elaborate ornaments or heavy weaponry. Ornaments come in all shapes and sizes, including colourful tail feathers and crests, often prominently displayed, while weapons include antlers, horns, and excessively long teeth – inevitably coupled with a willingness to use them on opponents. But, whether ornament or weapon, these additions are usually there for the same reason: to impress females and intimidate other males. On the other hand, the adult male and female of many species are almost impossible to tell apart. And many of these, like parrots and cockatoos, are hardly drab and dull – they are some of the most brightly coloured and beautiful birds in the world. So, what makes a bird lucky in love?

While males and females have equal success on average, the average doesn't tell the whole story. In some species, a few individuals of one sex – often the male – can potentially achieve extremely high fitness IF they employ the right tactics, leaving most males with nothing (and so restoring the average). It seems that gaining large numbers of mates and shouldering out the competition often involves extreme violence or dazzling displays, and so natural selection has favoured this behaviour, driving some species to greater and greater extremes.

To find out how this works, let's take a closer look at two birds with very different sex lives and investigate why such different strategies have evolved. We begin with the black grouse – a plump relative of the humble chicken that is found across the uplands of Northern Europe.

On a Scottish hillside just before dawn, a group of male black grouse have gathered for a performance. Dressed to impress in glossy black feathers set off by incredible scarlet eyebrows, they bow low and spread their tails to reveal an astonishing powder-puff of crisp white bum feathers. These, they proudly display to each other, all the while popping and twirling to make sure that everyone has had a good view. But this dance, although certainly impressive, is not performed in good spirit. Rival males are quick to show their displeasure, and rather than watching in quiet admiration, they attack with extended claws in the hope of reducing each other's costumes to a ragged pile of feathers.

Lurking in the vegetation at the side of the display ground is the cause of this boorish behaviour. The female black grouse is much less showy than the male, but beautiful in her own way – quietly dressed in gorgeous shades of brown and grey – and for the males on the showground, she is the only audience that matters. But she has not turned up to offer flattery. Instead, she is there to judge and soon decides on the winner. Moving onto the dancefloor, she mates with a single male, and having done so, she leaves. Over the next few days, other females will do the same – and strangely, they will nearly all pick the same male. For black grouse at least, it seems that some guys really do have all the luck.

The Only Audience That Matters . . .

Once she has mated, the female disappears to lay her eggs, and a few weeks later, her newly hatched chicks are taking their first steps through the heather. They are born in a state of advanced development and will soon begin to feed themselves and make tentative practice flights. But their mother is always close by, leading them to good feeding grounds and keeping an eye out for predators. In sharp contrast, dad continues his stint on the dancefloor, entirely oblivious to their existence. So, why does dad make such a hard-hearted decision?

Let's consider the options for a male black grouse. He can either abandon his first chicks in order to pursue more mates, or he can stay and help to rear them. We think that male black grouse choose to abandon the chicks because they are born in an advanced stage of development and will probably survive quite well without him. But, back on the dancefloor, the male can probably secure more mates and so become the father of even more chicks, and if he doesn't abandon the first ones, he will lose these opportunities.

The female black grouse is in a very different position, as she is literally left holding the baby. Bird eggs are some of the largest in the Animal Kingdom and females must invest resources from their own bodies in order to produce them. To abandon the chicks would mean wasting her investment, and so she cares for them and helps them to reach adulthood safely. Her costume is subdued because she doesn't have to attract a mate; instead, she will be vulnerable to predators while sitting on her eggs, and so needs to be camouflaged rather than flashy. It's easy to think that the males have the better deal, but a female black grouse is not powerless – very importantly, she can choose the father of her chicks. But if males don't help to rear the chicks, what exactly are the females looking for when they watch the dancers with such a critical eye?

This problem has puzzled biologists for years and there are no absolutely clear-cut answers. During their dance battles, males are trying hard to show off their health and vigour, so perhaps females are simply checking that the males are disease-free, to avoid catching something nasty during intimate contact. As well as her own health, a female might also be concerned for the health of her chicks, as they stand to inherit half of the male's genome. If a male is healthy enough to win the dance-off, there might be a better chance that he is carrying resistance to whichever version of the parasite is currently doing the rounds. So, by choosing the winning male, she could be choosing future health benefits for her offspring.

The extreme difference in the appearance and behaviour of male and female black grouse isn't the only way to organize a sex life – and indeed, is unusual among birds. So, let's walk down the hillside to investigate the alternative. On a calm brooding loch, two great crested grebes (elegant waterbirds) have taken a break from fishing to engage in a very different type of dance. Unlike the black grouse, one of the dancers is female, although it's almost impossible to tell the sexes apart. Sporting identical headdresses of puffed-out feathers, the two birds begin with a bobbing and weaving contest, where they mirror each other's movements and occasionally grab and wave pond weed – and if things go really well, they will rise up and dance on the surface of the water, kicking up spray with their webbed feet.

This dance is not designed to impress potential partners lurking on the sidelines. A pair of grebes dance only for each other, but they nevertheless subject each other to the severest scrutiny. Only if both are satisfied will they decide to mate – and if they do, then far from leaving his mate to go off and secure another partner, the male great-crested grebe helps to build the nest and takes his turn to incubate the eggs and feed their chicks. So, why are great crested grebes so passionate about sexual equality when black grouse seem determined to pursue outdated gender roles?

Unlike the precocious black grouse chicks, the chicks of the great crested grebes are helpless when born, and if one parent dies, they are unlikely to survive. There is little point in the male leaving his chicks to pursue other females, if by abandoning them, his first chicks are simply going to die. Instead, the male's best strategy is to commit to one partner – and the courtship dance allows him to see whether his potential mate is really the one for him. The female great crested grebe is looking for exactly the same qualities – a reliable male who will help to feed and rear her chicks, and the courtship dance allows her to see whether *he* is really the one for her. So, unlike black grouse, male and female great-crested grebes have near-identical sex lives, and they are almost impossible for human birdwatchers to tell apart.

Charles Darwin once said that the tail of the male peacock made him sick – he simply could not understand how natural selection could have produced a feature that so clearly reduced its owner's chances of survival. But peacocks are closely related to black grouse – their chicks are born in an advanced stage of development and the males don't help to rear them; instead, they invest their energy in elaborate costumes to gain as many mates as possible. Eventually, Darwin realized that there was a type of natural selection that he called *sexual selection*, which can drive the evolution of extreme characteristics when the stakes are high enough.

Sexual selection has led many animals to invest in elaborate structures and behaviours to allow them to attract and secure mates. In mammals, where males often invest little or nothing in their offspring, this can lead to extraordinary levels of violence – as seen in elephant seals – where males battle it out until bloody and exhausted. By securing a stretch of beach, elephant seal males can commandeer or persuade large numbers of females to mate with them, leaving other males out in the cold. But rather than succumb to zero fitness, the ousted males continually give battle to try to secure offspring of their own. As a result, males weigh up to three times

More than 500 fish species commonly change sex as adults

more than females and grow an extraordinary proboscis so that they can trumpet their magnificence all the way along the beach.

While the sight of a peacock's tail no longer makes biologists ill, one of the next great puzzles involving sex is exactly how the sex of individuals is determined. For most plants and some animals, like corals and snails, the question doesn't apply, as most of them are hermaphrodite. But if we return to the sparkling stream, the fertilised zebrafish egg cells are developing into fully functioning fish. Their sex will affect their development as males and females carry different sexual equipment and end up different sizes. So, how and when do the developing embryos find out whether they are male or female?

'Know thyself!' is a common philosophical cry. But, while all well and good in theory, self-knowledge is usually difficult to acquire, and sex is no exception. Some animals rely on their genomes for information and others on the environment. In mammals, biological sex (rather than gender) is genetically determined, and to understand how this system works, we need to look more closely at their chromosomes.

In chimps, one of the 24 pairs of chromosomes are called the *sex chromosomes* and they reveal the biological sex of their owner. In females, the two sex chromosomes look identical and are both called X, so females are designated XX. Just like all the other pairs of chromosomes, the two sex chromosomes are separated when eggs are formed inside the bodies of females, so each egg cell receives a single X chromosome.

In males, the two sex chromosomes look very different. One is an X chromosome, while the other is much shorter and called Y, so males are designated XY. When sperm cells are formed inside the bodies of males, the two sex chromosomes are also separated, but this time, half of the sperm cells will receive the X chromosome, and the other half will receive the Y chromosome.

When fertilization takes place, the sex of the new chimp baby is determined by the sperm cell when it enters the egg. All chimp eggs carry an X chromosome, so if the sperm cell also carries an X chromosome, then the new baby will be female, while if the sperm cell carries a Y chromosome, then the new baby will be male. Incidentally, there's no particular reason why males are XY and females XX. Birds do things the other way around: females have the oddball sex chromosome and males have two identical ones, although the reasons for this difference aren't known.

Sex determination in mammals and birds is really quite simple, and it's easy to think that other species must surely use the same sensible system. But, as with so many things in biology, nothing could be further from the truth. Zebrafish lack sex chromosomes and, despite having studied zebrafish embryos for decades, we still don't fully understand how their sex is determined. And some species don't rely on their genomes at all.

All crocodiles, most turtles and some lizards use signals from the environment to determine the sex of offspring. Females often bury their eggs in burrows or in a heap of vegetation and most are then left to develop alone (although crocodile mothers make excellent parents and guard the nest with violent intent towards intruders). The sex of the developing embryos is often, but not always, linked to temperature. In crocodiles, cooler temperatures favour the development of females, while warmer temperatures favour the development of males. This method often leads to populations with far more females than males, and it's still unclear why this happens, or what advantage, if any, it might bring.

One of the advantages of the sex chromosome system in mammals is that it naturally produces the same number of male and female offspring, because half the sperm cells carry an X chromosome and half carry a Y chromosome. This is important because a biased sex ratio is generally bad news. If a strange mutation arose that made males rare, then males would become hugely valuable as all females need a male to father their offspring. In turn, this would lead to enormous selection pressure to increase the frequency of males. Precisely the same argument can be made about a mutation that made females rare, and so natural selection has generally favoured mechanisms that ensure an equal sex ratio.

While biologists still pursue answers to many of the riddles posed by sex, some aspects are uncontroversial. We all agree that sex allows members of the same species to exchange parts of their genomes, creating a shared common genome that binds members of the same species together and allows useful mutations to be quickly passed around. Indeed, the most common definition of a species is a group of individuals that are able to have sex with each other. In contrast, members of different species generally do not have sex and so the shared common genome of each species is isolated from all others. This isolation is essential, or a new species couldn't evolve along its own path, and the Earth would be a much duller place.

Darwin's most famous contribution to biology is a book called *On the Origin of Species*. Widely read even today, the title is rather misleading, as the book is mostly focused on natural selection and how it produces organisms that appear to be perfectly adapted to their environments. But, while Darwin may have had less to say about how new species arise, modern knowledge of genomes has given us important new insights.

Hollywood producers love to imagine what might happen if two radically different animals successfully mated and produced viable offspring – Piranhaconda and Sharktopus being two particularly bizarre examples. Such hybrids are clearly science fiction, and to understand why these extraordinary creatures will never terrorize a swimming pool near you, let's consider the mule, which is a rather unusual domestic animal. Like all other animals, a mule has two parents, but *un*like most animals, they belong to different species – the mule has a horse for a mother and a donkey for a father. A mule is stronger than a donkey and more sure-footed than a horse, which is why humans continue to breed mules, especially in countries with mountainous terrain, but doing so presents certain difficulties. Mules are *sterile*, meaning that they cannot produce more mules, or indeed any offspring of their own.

The mule's sterility can be traced back to the double genome that sexual species possess. In most cells the genome is totally unaware of its double nature because every time a cell divides, the entire genome is simply copied and passed on to the two daughter cells. But, when egg or sperm cells are produced, they only receive one copy of the manual, so the genome has to be broken in half.

Unfortunately for the mule, separating its genome into two neat halves just doesn't work. A single copy of the horse genome contains 32 chromosomes, while a single copy of the donkey genome contains 31. To produce egg and sperm cells, the pairs of chromosomes have to line up and begin the process of recombination, when sections are swapped around to make new copies of the manual. But in the mule, the two copies of each chromosome are just too different for successful recombination, and one of the horse chromosomes (number 32) is left all by itself with no partner. These problems mean that egg and sperm cells simply cannot form properly, and so the mule – although a perfectly good animal in all other respects – is sterile.

If the horse and the donkey were wild animals, we would expect them to avoid romantic entanglements, as sterile offspring are a waste of precious resources. Indeed, we believe that the elaborate courtship rituals of many animals have partly evolved as a way to prevent individuals mating with the wrong species. This explains why courtship displays often involve key features unique to that particular species –

The Great Ape Family Tree

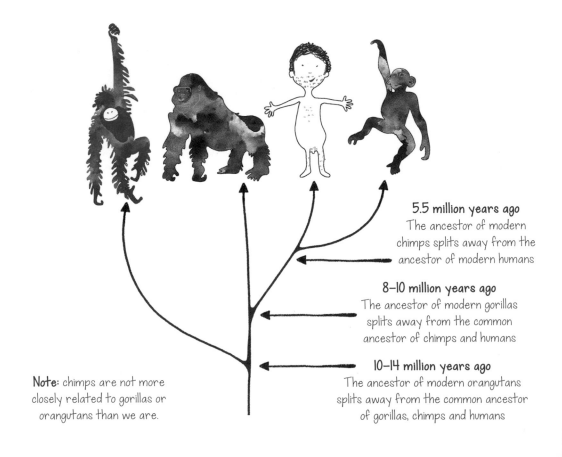

5.5 million years ago
The ancestor of modern chimps splits away from the ancestor of modern humans

8–10 million years ago
The ancestor of modern gorillas splits away from the common ancestor of chimps and humans

10–14 million years ago
The ancestor of modern orangutans splits away from the common ancestor of gorillas, chimps and humans

Note: chimps are not more closely related to gorillas or orangutans than we are.

blue-footed boobies (a type of seabird) hold up their blue feet for close inspection by potential partners, while parrots display their colourful plumage. But, if members of different species don't mate with each other, how do new species get started in the first place?

The commonest way for a new species to form is for a group to become isolated from the rest. This has happened frequently on islands, when a small group of animals have flown or rafted there accidentally, and been cut off from their mainland cousins. The conditions on islands are usually very different from the mainland, and so natural selection takes the new arrivals in a different direction – which is precisely why so many island birds have ended up flightless. But it's possible that the two groups might eventually find themselves back together in the same place.

On coming together, we might expect the two groups to welcome each other with open arms and begin the journey back to one species. But, if enough time has passed and their genomes have become too different, they simply might not recognize each other or the hybrid offspring – like the mule – might be sterile. Even if the offspring are fertile, they may still be at a disadvantage as they are likely to possess a strange combination of features from both parents that don't work well together, leading natural selection to cause them to start avoiding each other as potential mates. In the end, if the two groups are different enough, they will probably become separate species, and this is how the enormous diversity of life on Earth has gradually accumulated.

The longer two species remain isolated from each other, the more different their genomes become. Chimpanzees and humans are thought to have last shared a common genome around five million years ago, while we last shared a common genome with mice around 75 million years ago and with zebrafish, perhaps around 450 million years ago. We can estimate these dates partly because we have a fossil record, but partly because we have found ways to read the entire sequence of letters in the genome of any species and so compare them. The technique is called *genome sequencing* and it becomes easier and cheaper every day.

Genome sequencing can be used to find out when species last shared a common genome by comparing just how different their genomes are today. If the two genomes are similar, then it can't be very long since they split, but if they're very different, then they've probably been following separate evolutionary paths for much longer. Such comparisons are being used to build a Tree of Life: a grand enterprise in which we will be able to see exactly how all species on Earth are related to each other.

This project, carried out by thousands of scientists around the world, has already yielded many surprises. One example is a complete rethink of the ancestry of whales. Whales were once thought to have evolved from extinct carnivorous animals, but a comparison of their genomes with those of other living mammals came to the startling conclusion that their closest modern relative is the hippo. The delightfully named whippo hypothesis estimates that whales and hippos last shared a common genome around 55 million years ago, and although there is a good sequence of fossils that show how whales gradually became fully aquatic, the same can't be said for hippos, whose ancient past is shrouded in mystery and awaits further fossil finds.

Comparisons can also reveal our own ancestry. While Darwin had suggested that humans were closely related to the great apes (a suggestion that shocked many Victorians), the precise identity of our nearest relative was still being debated as late as the 1980s. Both the gorilla and the chimpanzee were in the running, but a comparison of genomes has made it clear that the chimpanzee is our closest living relative, with a genome that is 98% similar to ours. We have to be careful about what such comparisons mean – it has been claimed that a typical human and a typical chimpanzee are genetically more similar to each other than two randomly selected fruit-flies of the same species – but we can all agree that the small genetic differences between chimpanzees and humans have profound consequences for both species.

While sexual reproduction has been crucial to the evolution of multicellular creatures, for most of Earth's history, cells lived alone. Yet in order to process and transmit information effectively – and consequently evolve – they needed a source of energy and raw materials. But cells are up against the laws of the universe, and need to comply with all of their demands.

Chapter 4
ENERGY
Better living through chemistry

In our universe, the laws of thermodynamics govern the behaviour of energy and matter, making it an orderly and predictable place. All things are subject to their iron rule, but life has pushed them to the limits, leading generations of humans to believe that there must be some special animating force – perhaps even a divine spark – that sets living things apart. But although living things can indeed appear miraculous, they cannot and do not break the rules, and their undoubted spark is entirely a product of physics and chemistry, set alight by natural selection.

The first law of thermodynamics is uncontroversial: matter can neither be created nor destroyed, but to understand how life's magic can be confused with sorcery we need to look more closely at the second law. The *second law of thermodynamics* concerns itself with *entropy*, which can be roughly translated as disorder, and the law states that the entropy of a closed system must always increase, as a simple example demonstrates.

Imagine a pile of neatly stacked Lego bricks on a table. If the bricks were sent tumbling to the floor, they could end up strewn around in myriad ways, none of them particularly surprising. Surveying the mess, we might feel annoyed, but we're unlikely to leap to the conclusion that there's something wrong with the world, as we have simply witnessed the second law in action. Entropy is disorder, and most of us have learned from bitter experience that disorder always increases, even if we didn't know that this was written into the fabric of the universe.

Now imagine that, as the pieces fell to the floor, they spontaneously assembled themselves into a boat or a house. Under the second law, such an event is so fantastically improbable, that we might either think that we were dreaming or that we had wandered onto the set of a Harry Potter film. Turning our attention to life, we can see why some have tried to claim that cells are the living equivalents of spontaneously assembling Lego sets. Cells build large proteins from thousands of amino acids and enormous genomes from billions of tiny nucleotides, and eventually they have enough of these molecules to construct a brand-new cell. So surely, this must cause a decrease in entropy, in defiance of the second law? But if we look more closely, we can see why cells do not require access to supernatural help.

No one would be surprised if a child came along and used the bricks to build a

The Second Law of Thermodynamics

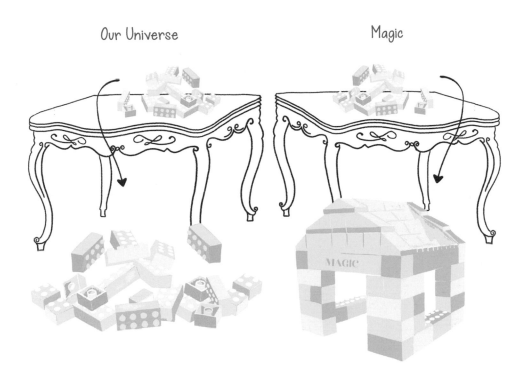

Our Universe   Magic

house or a boat, and unless they had built something truly incredible, most of us wouldn't brand our child miraculous for doing so. The child hasn't flouted the second law, because to build the house they relied on muscle power, and fuelling muscles requires an input of energy. This distinction is crucial because the second law only applies to a *closed* system, and all living things are intimately connected to their surroundings.

In the case of the child, they will only have the energy to build the Lego house if they have been given regular access to food. Given what we know about human children, perhaps they have recently consumed some of their favourite ordered structures, like chicken, chips or cheese. Inside the child's gut, these will have been broken down into their constituent molecules, which are needed by the muscle cells to do work on the Lego house. It's pretty clear that this process caused the chicken, chips or cheese to become vastly more disordered, and this compensates for the increase in order caused by building the Lego house. So, a child can create order in one place and *not* defy the second law, by creating a lot more disorder somewhere else (demolishing chicken, chips or cheese, according to preference). And a cell too can build large molecules – and even new cells – by harnessing an external source of energy and so disordering their surroundings.

Most cells obtain the energy they need to build and repair themselves from a very limited number of sources. The ultimate source of energy on our planet is the Sun, which plant cells directly harness, while animal and fungal cells rely on chemical energy that comes from food. But the earliest cells on our planet were much more imaginative. They belonged to two different groups called *Bacteria* and *Archaea* (pronounced: *Are-kay-ah*) – both chemical wizards – that invented all kinds of ways to extract energy from unlikely sources. Indeed, plants, animals and fungi only arose because they stole one or two recipes from these ancient alchemists and turned them into a multicellular show-stopping feast.

An active volcano – Mount Kirishima – looms over the surrounding landscape of the Kirishima National Park in Japan. Rooted in the deep layers of the Earth, the volcano channels molten rock to the surface, feeding a series of hot springs and muddy ponds scattered around its base. These waters are not for the fainthearted: within each steaming pot is a chemical brew whose combined heat and acidity would quickly destroy any plant or animal, but for some life on our planet this is an ideal playground.

Living in the springs, single-celled organisms belonging to the group Archaea are pushing the envelope of life's possibilities. The name means 'ancient ones', but although they have been around for at least 3.8 billion years, so have bacteria, so their name is somewhat misleading. Despite this venerable history, biologists only realized that Archaea were a separate branch of life in the 1970s. To the untrained eye, bacteria and archaea certainly appear to have a great deal in common: most of them are small cells with rigid shapes that live alone and do not form larger bodies; but crucial differences in their membranes, their ribosomes and their cell walls mark them out as two distinctive groups.

Around five million trillion trillion bacteria live on today's Earth – that's the number five followed by 30 zeroes – far outstripping the total count of plants and animals and even eclipsing the number of stars in the known universe. Archaea are rather less common. We used to think that they could only be found in extreme places where no other life could scrape together a living, but we now know that archaea are abundant in the oceans and in soil, where their versatility is essential for the growth of other organisms.

If bacteria and archaea did indeed arise in deep-sea vents, then the first step on their journey to conquering the rest of the planet was to move out into the wider world. Some vents are chemically active, and could have provided the earliest cells with essential building blocks, like amino acids, and a source of energy. But the world beyond the vents would have been barren and empty, meaning that cells finally needed to go it alone.

The solutions to the twin problems of generating energy and constructing building blocks are deeply intertwined. Both required cells to master the rules of chemistry, which, in turn, meant delving inside atoms and making them work to their advantage.

All matter is made from *atoms*. Atoms are so tiny that we used to think they were indivisible – the smallest possible particles in the universe – but we know now that atoms are constructed from even tinier parts. In the centre of an atom is a tight cluster of *protons* and *neutrons*, while tiny *electrons* whizz around them, like planets orbiting the Sun. Each atom contains an equal number of positively charged protons and negatively charged electrons, so the atom itself is neutral. This slightly dull arrangement may seem an unlikely starting point for building a universe, but we shouldn't be deceived by the apparent simplicity of atoms: they exhibit an enormous range of behaviour, depending on which *element* they belong to.

There are around 91 naturally occurring elements on Earth, each defined by the number of protons they carry. Hydrogen, with its single proton, is the lightest element, while inside the nucleus of a uranium atom, 92 protons jostle together uncomfortably. Mediaeval alchemists dreamed of transforming one element into another, but we now know that changing the number of protons inside an atom releases energy on a truly frightening scale, and until the advent of nuclear physicists, living things had the sense not to mess with it. But, while protons are difficult to shift, the electrons circling the nucleus can more easily be persuaded to move around.

Within atoms, the orbiting electrons are arranged in a series of shells. Each shell has a maximum capacity – for example, only two electrons can be held in the shell closest to the nucleus. If all of the shells in an atom's possession are filled to capacity with electrons, then these electrons will not be tempted to leave their parent atom. We know this, because a group of elements, called the noble gases, which include helium, neon and krypton, naturally occur in this happy state, and as a consequence they are notoriously unreactive. Indeed, the noble gases inhabit a kind of atomic bliss – a state which other elements are very keen to reach.

The shells of most other elements are not filled to capacity with electrons. But by losing, gaining or sharing electrons with other atoms, many elements can come close to this hallowed state. Indeed, the reshuffling of electrons among atoms is the basis of all chemistry and allows new chemical compounds and molecules to form. Better still, these rearrangements often liberate energy, which means that cells can harness chemical reactions to provide the energy they need to offset the decrease in entropy that their activities would otherwise cause. Harnessing chemical reactions to do battle with the forces of nature is something that humans have also been keen to try out. In fact, chemistry has famously allowed a small number of people to defy – not the second law of thermodynamics – but gravity.

In July 1969, three men were sitting inside the top of a Saturn V rocket on the launchpad at the Kennedy Space Centre. Their destination was the Moon – and the world was watching. Taller than Big Ben and three times heavier than the London Eye, Saturn V was by far the largest rocket that the world had ever seen. But most of it wasn't dedicated to carrying sophisticated equipment, lunar rovers or crowds of excitable Moon-tourists. Indeed, most of it wasn't going to the Moon at all.

The Saturn V rocket had four parts. Only the uppermost section would leave the Earth's atmosphere and journey to the Moon, carrying three astronauts and all the equipment they needed to orbit the Moon, land there, take off again, and return to Earth. The remaining three parts – and 85% of the total weight – were just tanks

Chemistry is Driven by the Behaviour of Electrons

Helium is a noble gas with just two electrons. Its outer electron shell is full, making helium smug (and unreactive)

Neon is a noble gas with ten electrons. Like helium, its outer electron shell is full, so neon is also smug (and unreactive)

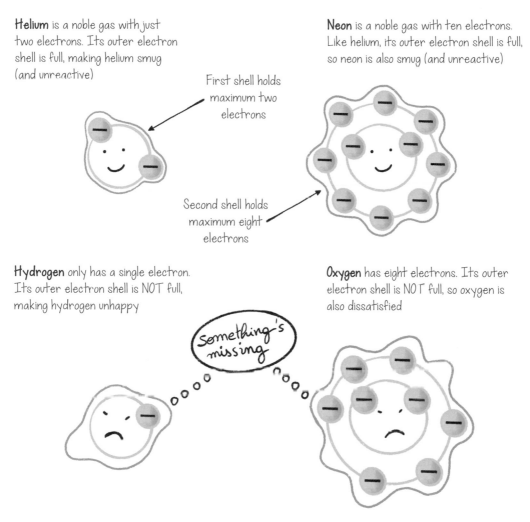

First shell holds maximum two electrons

Second shell holds maximum eight electrons

Hydrogen only has a single electron. Its outer electron shell is NOT full, making hydrogen unhappy

Oxygen has eight electrons. Its outer electron shell is NOT full, so oxygen is also dissatisfied

Something's missing

By sharing their electrons and forming a molecule of water, two hydrogen atoms and one oxygen atom can attain 'noble-gashood' and have full outer shells. All three atoms are now happy, so water is a very stable molecule

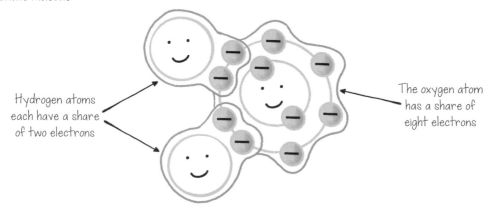

Hydrogen atoms each have a share of two electrons

The oxygen atom has a share of eight electrons

filled with the stuff required to fire the rocket off the launchpad and escape the Earth's gravity, and would be jettisoned within minutes of launch.

To generate the kind of lift needed to move thousands of tonnes of metal into space, the scientists at NASA turned to chemistry. Saturn V was carrying enormous volumes of liquid hydrogen and liquid oxygen, carefully stored in separate tanks. When these two elements are brought together, the resulting reaction is so violent that it can lift the rocket off the ground, in a spectacular triumph of chemistry over physics. Interestingly, no one watching claimed that they were witnessing a miracle – just good-old human ingenuity – but what exactly was going on once the countdown had ended and the rocket began to inch its way past the supporting tower on its way to the moon?

Hydrogen is a fuel, and burning any fuel is a chemical reaction. All reactions involve electron rearrangements, and in this case, hydrogen and oxygen team up to share their electrons and so move closer to a state of noble-gashood. The result is the formation of a new molecule, called water, with the familiar formula H_2O. The formula reveals that each water molecule contains two hydrogen atoms and one oxygen atom, tightly bonded together by the electrons that they now share. And because the electrons are now happier (officially in a state of lower energy), the formation of water is accompanied by the release of a massive amount of energy.

If a Saturn V rocket had exploded on the launchpad, it would have been the equivalent of detonating around 500 tonnes of TNT. To put this in perspective, a similarly sized explosion took place on 11 August 2020 in the port of Beirut in Lebanon. This was also caused by a chemical reaction – 2,750 tonnes of ammonium nitrate had been unsafely stored in a warehouse – and the explosion devastated the port, creating a crater 140 metres wide and hurling ships into the air, leaving 190 people dead.

But not all chemicals are inherently unsafe, nor do all chemical reactions have the power to devastate their surroundings. Hydrogen is an excellent fuel and burns easily because, by sharing their electrons and forming water, oxygen and hydrogen can both attain their dreams of atomic bliss. Indeed, burning hydrogen is an example of a *spontaneous reaction* – metaphorically speaking, we need only apply a spark and then move rapidly backwards and admire the fireworks. Spontaneous reactions are the equivalent of rolling a rock downhill or pushing against an open door, and this type of reaction very often releases energy that can be harnessed to do work.

Unfortunately, many of the reactions that cells want to carry out – like the joining together of nucleotides to make a long DNA strand – are not spontaneous, and instead, energy will have to be invested to make them happen. But just as Saturn V could defy gravity by burning enough fuel, so a cell can carry out reactions that are the equivalent of pushing rocks uphill IF it can harness enough energy from a suitable spontaneous reaction. It's all a delicate balancing act and cells have achieved it in much more subtle and interesting ways than simply setting fire to a tank of hydrogen. Indeed, if cells were asked to comment on our efforts, they might say that in comparison with their sophisticated strategies, our puny efforts on the launchpad are simply sound and fury, signifying nothing.

Inside a cell there are thousands of energy-hungry activities, like assembling RNA messages or joining up amino acids to form new protein chains. These activities take place at different locations, so energy needs to be delivered to multiple places at the same time. In a modern house we have a similar problem – the kettle, the computer and the television all need a power supply – and our solution is to install a network of wires that deliver electricity throughout our homes. Very few of us actually generate our own electricity, although some of us may have mounted solar panels on the roof; instead, it is generated in power stations that are then hooked up to the National Grid for efficient distribution.

Amazingly, cells have hit upon a rather similar solution. The myriad activities that go on within a cell are powered, not by electricity, but by a small molecule called *ATP (adenosine triphosphate)* that acts like a one-size-fits-all battery. ATP batteries can be plugged in anywhere and everywhere, and cells certainly run through them at an astonishing rate. A typical human will use their own bodyweight in ATP every single day, so recharging the batteries must be done quickly and efficiently, millions of times every day.

The power stations that act as battery-recharging sites are located within cell membranes. Made from a double layer of a fatty substance called *lipid*, membranes are used to form a boundary between the cell and the outside world, although plant and animal cells also use internal membranes to divide the cell up into smaller compartments. Very small molecules, like oxygen and water, can move freely across cell membranes, but larger molecules or those that carry an electrical charge have to enter through specialized channels, allowing the cell to control their entry and exit. And these channels are crucial for the proper operation of the cell's power stations.

To understand how the cell's power stations work, let's take a closer look at a power station built by humans. Inside a power station, electricity is generated by turning turbines, and to do so means harnessing a source of energy, like the wind or running water. In a hydro-electric power station, the turbines are turned by water flowing downhill, and the power station is often built into a dam that holds the water back, allowing the flow of water through the turbines to be precisely controlled.

The cell's power stations are modelled along exactly the same lines. At the business end of the power station, massive molecular machines that act as turbines pierce the membrane from one side to the other. But instead of generating electricity, these machines are hooked up to the dead ATP batteries and they are energized by the flow of positively charged protons, rather than by running water. As protons flow through the miniature turbines, ATP batteries are recharged, and they spin fast enough to recharge more than 100 batteries every second. These batteries are then released and used to drive reluctant chemical reactions all over the cell. Simple, but brilliant.

Of course, there's just one problem. To generate the flow through the turbines, masses of protons must be herded together on one side of the membrane where they build up, ready to flow back and recharge the cell's batteries. But the second law of thermodynamics always demands a maximum state of disorder, so piling up molecules in one place is in flagrant defiance of the law. So now we have come to the crux of the matter – how were the protons piled up in the first place?

If we look along the membrane, we see that the turbines aren't the only machines present. There are also equally impressive-looking pumps that push protons back across the membrane. But piling up protons in one place is like pushing water uphill, so it's these proton pumps that need to be hooked up to an external source of energy.

Incredibly, in all cells, the energy needed to drive the proton pumps comes from electrons. Bacteria and archaea have found many different ways to obtain them, creatively harnessing electrons from unlikely chemical reactions involving elements like iron and sulphur. Although these reactions keep cells alive, they generally don't yield enough energy to support a high-octane lifestyle, but around 2.4 billion years ago, a crucial event in the Earth's history filled the atmosphere with oxygen. And the presence of this electron-hungry molecule paved the way for a super-charged biosphere, by allowing cells to harness the same reaction that NASA used in the design of Saturn V.

In animal cells like ours, hydrogen is shuttled to the power stations by a special carrier molecule and separated into a proton and an electron. The electrons are then tempted away by offering them a better home, but as they jump, they are forced to do work for the cell. The energy they release is used to fire up the pistons and pump protons to the other side of the membrane; but this isn't one long glorious burn. Instead, each electron will pass through three different pumps, and at each stage it loses some of its energy by activating the pump. Only at the third stage, with their energy spent, are the electrons reunited with the protons, as they return through the turbines. The resulting hydrogen then combines with oxygen to form water.

Proton pumping at membranes is an ancient feature of cellular life, shared by bacteria, archaea and ourselves. Indeed, animal and plants cells stole this technology from bacteria (as we will discover in a later chapter) allowing animals at least to adopt hyperactive lifestyles. So, if life has a spark – then this is it. Protons are charged particles and the endless proton pumping means that cells maintain a voltage across their membranes that if scaled up to one-metre thick, would unleash the kind of energy seen during a lightning strike. It is this energy that fuels everything else in the cell, allowing life to appear miraculous and even providing the power to keep enormous collectives of co-operating cells walking and talking.

Finally, we might reasonably ask where cells today obtain a source of hydrogen. Hydrogen is the lightest element and hydrogen gas easily escapes into space, so there is rather little in the Earth's atmosphere. But cells are full of hydrogen because it is bound up inside many common molecules. To be useful at the cell's power stations, the right molecule has to be carefully chosen. The key to liberating energy is to find a

ATP Batteries are Recharged at Cell Membranes

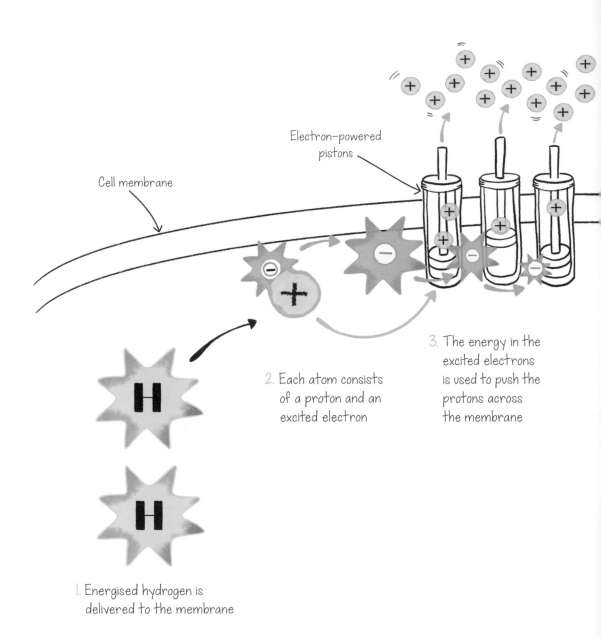

Electron-powered pistons

Cell membrane

1. Energised hydrogen is delivered to the membrane

2. Each atom consists of a proton and an excited electron

3. The energy in the excited electrons is used to push the protons across the membrane

Piled-up protons

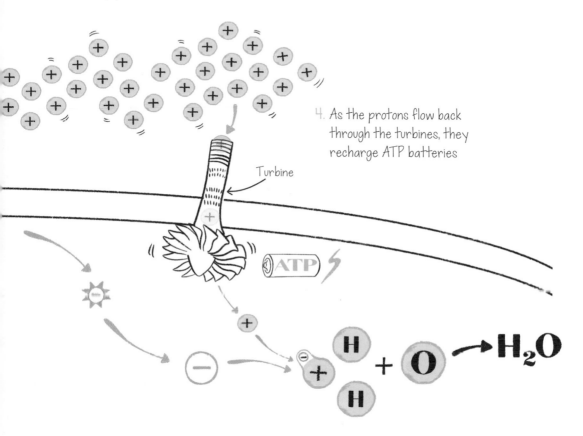

4. As the protons flow back through the turbines, they recharge ATP batteries

Turbine

ATP

H₂O

5. The energy from the electrons is now spent. The protons and electrons recombine to form hydrogen, which reacts with oxygen to form water

hydrogen-containing molecule that can be broken up without having to invest a huge amount of energy – so water, for example, in which hydrogen has found a very happy home, would be out of the question.

To find out which molecules have been chosen by animal and plant cells, we need to first take a look at the other side of cellular chemistry. As well as a source of energy, cells need building blocks for growth and repair, and they are all based on one element that has incredible versatility.

Let's begin by considering why cells need so many new building blocks. A typical human is a good place to start. The cells in a human body have a limited lifespan and need to be regularly replaced – white blood cells that fight invading bacteria only give battle for around thirteen days before succumbing to wear and tear; skin cells last for around one month before being sloughed off, contributing to the dust under the bed; and red blood cells carry oxygen for around four months before they too give their last gasp. To compensate for these ongoing losses, a human body needs to make a few million new skin cells every day and a staggering 2.5 million new red blood cells every second.

New cells require new molecules, and all of the cell's building blocks are constructed from *carbon*, an element which is peculiarly well-suited to the job. A carbon atom has enough space to share up to four additional electrons with other atoms, allowing it to make a dazzling array of *organic molecules*, which are the basis of all life. Some, like methane (CH_4), are very small, but there appears to be no limit to the size of molecules that can be formed when carbon atoms gang together. No other element has quite the same versatility, which means that alien life, should we eventually find it, will almost certainly be carbon-based too.

There are four types of carbon-based building block that cells require, and each type can link up to form much larger *macromolecules*. The first building blocks are simple sugars, like *glucose* or fructose, and they contain just three elements: carbon, hydrogen and oxygen. Simple sugars can be strung together to make enormous macromolecules, like *cellulose*, which strengthens plant cell walls. Together, simple sugars and their macromolecules are known as *carbohydrates*. The name means hydrated carbon, because within most carbohydrate molecules there are two hydrogen atoms for each oxygen atom, as found in water (H_2O).

The second type of building block are the fatty acids, which are also made from carbon, hydrogen and oxygen. They are used to construct cell membranes, which require a special lipid molecule that looks like a tadpole with a water-hating tail and a water-loving head. This love/hate relationship means that lipid tadpoles spontaneously form a membrane that consists of two layers of tadpoles, arranged so that their water-hating tails are buried on the inside and their water-loving heads facing outwards. Excess fatty acids are stored in the body as globules of fat.

The third type of building block are the 20 amino acids from which proteins are formed. The amino acids are diverse, but all require the element nitrogen as well as carbon, oxygen and hydrogen, and two amino acids (methionine and cysteine) also require the element sulphur. Amino acids can be joined together in any sequence to form a protein, and sometimes carbohydrate molecules can also be attached, making a *glycoprotein*. A well-known example of a glycoprotein are the *antibodies* produced by our immune systems. These bind to invading viruses to prevent them from entering cells and wreaking havoc.

The final type of building block are the nucleotides, from which long strands of DNA and RNA are made. Nucleotides contain the elements nitrogen and phosphorus, in addition to the standard carbon, hydrogen and oxygen. As well as forming the individual letters of genomes and RNA messages, the ATP batteries on which cells depend are just an adapted form of one of the nucleotides (the letter A).

Animals need to obtain most of their building blocks by eating plants or other animals. In the UK, the National Health Service recommends that adults eat around 50 grams of protein, 70 grams of fat and 260 grams of carbohydrates each day – although this balance doesn't have to be too exact. In our digestive systems, the large molecules from which the unfortunate plants and animals were made are broken down into their building blocks, which are then absorbed, transported and delivered to all the cells in the body. On delivery, they are taken up and built back into human proteins or human DNA, but linking building blocks together is an

The **fastest** enzymes can process a million molecules per second

energy-demanding process. So, which of the building blocks can be used to liberate energy at the cell's power stations?

Each of the building blocks contains hydrogen, so it seems that any of them could be used to pump protons at the cell's power stations and so provide the cell with the energy it needs. But, in reality, the best molecule for this purpose is glucose, which is why our recommended diet is so carbohydrate-heavy and why many of us crave sugar when we're tired. Glucose is the molecule of choice because it can be broken down easily and quickly to release the hydrogen that it contains (and our brains will use nothing else). The hydrogen is shuttled away by a special carrier molecule and, on arrival at the power station, the electrons activate the proton pumps as they jump from the carrier molecule to oxygen, where they form water.

To release the hydrogen and so liberate the energy that animal cells require, glucose will have to be dismembered. This dismemberment also releases carbon dioxide, because the glucose molecule contains carbon atoms, as well as hydrogen atoms, and the first law of thermodynamics tells us that atoms can't be destroyed – only moved around and reshuffled.

The dismemberment of glucose happens gradually and requires many different small steps. Nearly every step is a spontaneous reaction – the equivalent of rolling a rock downhill – so we might think that the process will happen easily and quickly. But, unfortunately, it's not that simple, because even spontaneous reactions can be slow and need to be speeded up. Enzymes can massively reduce the so-called *activation energy* needed to get a reaction going by bending and twisting molecules in just the right way. So, the right enzymes allow cells to dismember glucose at low temperatures and at lightning speed.

As the glucose is gradually dismembered, new molecules are created, and these can be used to produce new building blocks, as well as to liberate energy. So, if the cell has plenty of glucose but is short of fat, then some of the new molecules can be siphoned off to make fatty acids. Or, if the cell is short of glucose, fatty acids can be used to liberate energy instead, which is why animal bodies store fat. Even leftover amino acids can be used to produce energy, and this versatility means that animals can adapt to a changing diet while still having the right balance of building blocks

The Dismemberment of Glucose

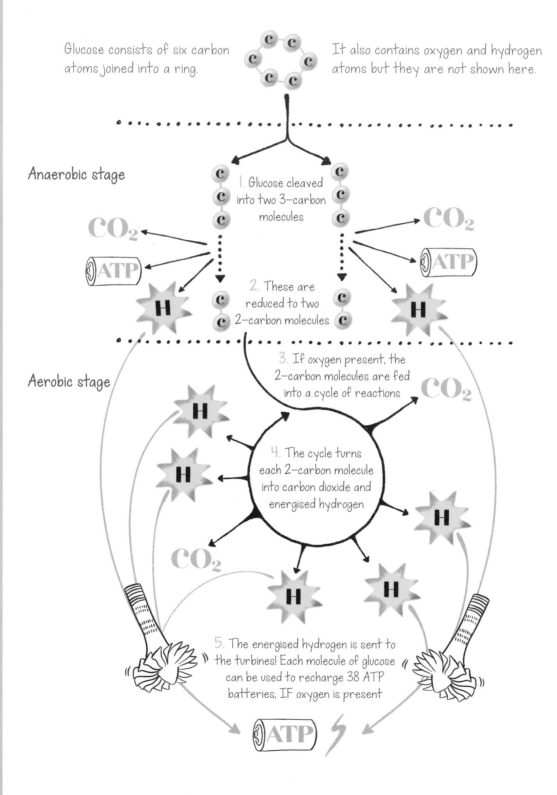

Glucose consists of six carbon atoms joined into a ring.

It also contains oxygen and hydrogen atoms but they are not shown here.

Anaerobic stage

1. Glucose cleaved into two 3-carbon molecules

CO_2

ATP

2. These are reduced to two 2-carbon molecules

H

CO_2

ATP

H

Aerobic stage

3. If oxygen present, the 2-carbon molecules are fed into a cycle of reactions

CO_2

4. The cycle turns each 2-carbon molecule into carbon dioxide and energised hydrogen

H

H

H

CO_2

H

H

5. The energised hydrogen is sent to the turbines! Each molecule of glucose can be used to recharge 38 ATP batteries, IF oxygen is present

ATP

and fuel, and there's no need to worry too much about the exact proportions of the different food groups that we eat.

When oxygen is freely available, the dismemberment of a single molecule of glucose allows an astonishing 38 ATP batteries to be recharged, but only 36 of these are recharged using the turbines in the cell's power stations. The first two batteries are recharged using a different method that doesn't require oxygen, allowing human muscles to keep working for a short time if oxygen just can't be delivered to them fast enough. This inefficient process is called *fermentation* (also known as *anaerobic respiration*) and many bacteria and archaea rely on it entirely, allowing them to thrive in environments from which oxygen is permanently absent. In humans, fermentation is only short-lived – it releases a toxic product called lactic acid – and many of us have probably experienced the burning sensation it creates in our muscles when forced to do too many star jumps by over-enthusiastic PE teachers.

The process of liberating energy from molecules like glucose in the presence of oxygen is called *aerobic respiration*. This extraordinary process squeezes every last drop of energy out of each molecule of sugar and is only possible because of the attack-dog nature of oxygen and its insatiable greed for electrons. If there was no oxygen in the atmosphere, then animals might have to rely on fermentation or some other poor substitute for aerobic respiration, and it's impossible to imagine that this could support the energetic lifestyles that we enjoy. So, without oxygen there would be no cheetah running by at 100 kilometres per hour and no kangaroo leaping more than seven metres in a single bound.

Animal cells need to undertake reactions that are energy-demanding, such as building new proteins from strings of amino acids, and they do this by harnessing the power of aerobic respiration to recharge tiny ATP batteries. By dismembering glucose, they can obtain both the fuel and the building blocks that they need – through the somewhat underhand method of eating other cells and breaking down their organic molecules – but they would quickly grind to a halt without the oxygen that lurks in the atmosphere.

Animal cells will starve if they have no organic molecules to eat, and yet they thoughtlessly destroy glucose every day just to stay alive. Worse, in doing so, they convert life-giving oxygen into water, and so remove two precious molecules from the biosphere with every breath they take. So, how on Earth do animals persist?

Animals are lucky that other living things have chemical tricks that they do not possess. On today's planet much of the oxygen in our atmosphere is produced by plants, but they certainly weren't the original source of atmospheric oxygen. Plants arrived late to the oxygen-producing party, and their arrival was only made possible because they stole what is arguably the greatest invention of all time from a humble bacterium: the ability to harness the sun's energy to make organic molecules, like carbohydrates, from scratch.

Plants don't eat, so they can't obtain their building blocks from other living things. But there is an alternative source of carbon on our planet – the gas carbon dioxide (CO_2) – and animals release six molecules of carbon dioxide for every glucose molecule that they destroy. The clue to how plants use carbon dioxide lies in a reaction we've already seen – aerobic respiration – where glucose is converted into carbon dioxide and water with the help of oxygen. In theory, all chemical reactions can be reversed – so carbon dioxide and water could potentially be converted into glucose and oxygen, allowing carbohydrates to be pulled essentially out of thin air.

There's just one problem with this proposed wizardry. Reactions might be reversible, but they occur spontaneously and release energy in one direction only. So, if aerobic respiration is the equivalent of rolling a rock down the slopes of Mount Everest, then reversing it requires pushing that enormous rock all the way back up to the summit from base camp. But, while no energy source here on Earth is big enough and safe enough for cells to harness to undertake this Sisyphean task, there is an extra-terrestrial body that can pack the necessary punch.

The ultimate source of nearly all the energy on our planet is the Sun. The Sun seems rather benign from Earth as it gives off warmth in a fantastically reliable way and most of us enjoy it. But it's 92 million miles away, and at close range, the Sun looks a lot less friendly. The Sun's surface temperature is around 5,500°C, and this heat isn't generated by messing about with chemistry – instead, the Sun indulges in hard-core nuclear physics. Its atom-smashing antics release energy on an entirely different scale to that seen when electrons get excited: radiation pours forth from the surface of the Sun – releasing one million times more energy in a single second than the humans on our planet consume in a single year – and certainly more than enough to fire up a reluctant chemical reaction.

Back here on Earth, any cell wanting to make glucose from carbon dioxide and water has a serious problem. Making a molecule of glucose requires breaking up water molecules to extract the hydrogen atoms, which can then be attached to the carbon dioxide. And this means prising electrons away from the water molecule in which they have found a safe and happy home.

Around two to three billion years ago (the exact date is still hotly disputed), a group of bacteria called *cyanobacteria* evolved a complex of molecules called a *photosystem* to smash up water molecules and so release hydrogen. The photosystem is a fiendishly complicated structure involving multiple proteins bonded together and embedded in a membrane. The job of most of the photosystem is to harvest light from the Sun and shine it into the heart of the system, where a molecule called *chlorophyll* lies. Absorbing light excites the chlorophyll to such an extent that it fires off an electron, which is trapped by the cell, leaving the chlorophyll one electron short. The desire to replace this electron is so strong that the bereft chlorophyll rips an electron out of a molecule of water to restore its normal state, leaving a shattered water molecule behind. The hydrogen released is then shuttled away where it is used to build sugar from carbon dioxide, with the help of ATP batteries (although the details of this process are fiendishly complicated).

The process of manufacturing sugar from carbon dioxide and water by harnessing the energy in sunlight is called *photosynthesis*. Its success relies on the excitable chlorophyll molecule at the heart of the photosystem, which we can thank for making the world green. The visible light pouring out of the sun comes in many colours – revealed when a rainbow decorates the sky. Chlorophyll absorbs some of these colours and with it the energy to fire off electrons, but in one of the great unsolved mysteries of the world, green light is useless in this process and simply bounces off the leaf unused. To a human admiring the world, this reflected green light is all we

Photosynthesis

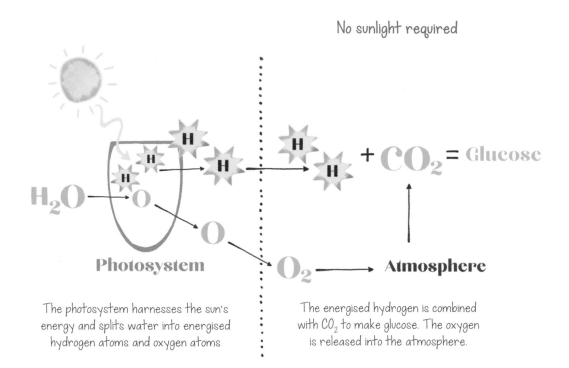

No sunlight required

H_2O

Photosystem

$+ CO_2 =$ Glucose

$O_2 \longrightarrow$ Atmosphere

The photosystem harnesses the sun's energy and splits water into energised hydrogen atoms and oxygen atoms

The energised hydrogen is combined with CO_2 to make glucose. The oxygen is released into the atmosphere.

see, but if chlorophyll could use all the colours in the rainbow, then no colours would be reflected back and the leaves around us would look black.

When the sun shines, plants can also use the photosystem to pump protons and recharge their ATP batteries, but these only work during the day. So, plants also carry out aerobic respiration and use some of their precious glucose to provide energy to keep other reactions going. Once glucose is produced, plants can use it as the starting point to make all the other building blocks they need, although they will need other elements like nitrogen and phosphorus to make amino acids and nucleotides, and they take these up from the soil.

The photosystems are brutal machines, so spare a thought for the poor oxygen atom, left stranded, when the hydrogen was so rudely ripped away during the destruction of water. Oxygen atoms can't stand being alone and so, as water molecules are pulled apart by the photosystems, they pair up and drift away as molecules of O_2, unloved and unwanted by the cell that generated them – and possibly straight into the lungs of a nearby animal that is keen to reverse the process that caused them to form in the first place. If there is such a thing as true harmony in the world, then this is it – the interplay between all the world's living things as they generate and consume oxygen in such a way to keep current levels of atmospheric oxygen steady at around 20%.

But this balance hasn't always been present. In truth, photosynthesis was one of the most disruptive innovations that our planet has ever seen. It marked a turning point for life on Earth, and by poisoning the atmosphere with oxygen, spelled doom for many of the poor bacteria and archaea that had made the planet habitable in the first place.

The very young Earth was so hostile to life that we call the first phase of its history the *Hadean*, after the Greek word for hell. Battered by fragments left over from the birth of the solar system, the surface would regularly melt over large areas, creating conditions that even the toughest life form would find impossible. But around four billion years ago, the Earth's surface cooled, allowing the first oceans to form, and we think that life then followed rather quickly

We have seen that life might have emerged in volcanic vents deep in the ancient oceans. Here, the earliest cells could have harnessed natural proton gradients to turn a primitive turbine, providing them with ATP batteries that could fuel the synthesis of new molecules. But cells couldn't break free of their birthplace until they could pump protons for themselves, and this required a source of energy. Certainly, aerobic respiration was out of the question as the atmosphere of the early Earth was entirely devoid of oxygen, and glucose would have been equally impossible to find.

As well as needing a source of energy, the first cells had to make organic molecules from scratch. Carbon dioxide is the obvious starting point, and the early atmosphere was full of it – but to turn carbon dioxide into glucose requires a great deal of energy and a source of hydrogen. The photosystems invented by cyanobacteria are the perfect solution to both these problems, but they are supremely complicated and must have evolved from something simpler.

Amazingly, there are still bacteria on today's planet that use simplified photosystems to harness the sun's energy. None of these are powerful enough to break up water molecules, so they don't produce oxygen as a waste product, but there are other molecules which aren't quite so challenging to smash up. One molecule that can be broken apart more easily is *hydrogen sulphide* (H_2S), and there is a bacterial photosystem that has evolved to do just that. Hydrogen sulphide (H_2S) is easier to destroy than water (H_2O) because the sulphur atom is a comparative weakling, and has a much less powerful grip on the hydrogen atoms than does the oxygen atom. And although hydrogen sulphide isn't a common molecule on our planet, it's abundant around areas of volcanic activity, and by exploiting it, these bacteria unintentionally set the stage for something better.

Once a photosystem had evolved that could pull hydrogen sulphide apart, then the machinery to destroy water was just a few evolutionary steps away. Water is so much more plentiful than hydrogen sulphide that any organism pulling off this trick could truly conquer the world, and in the end, the cyanobacteria managed it. They used the ingenious trick of rigging together two simpler photosystems to create a monstrous machine with sufficient power to drag electrons from water, and their unwanted waste – molecular oxygen (O_2) – began to trickle into the atmosphere.

Around 2.4 billion years ago, the levels of oxygen in the Earth's atmosphere suddenly jumped up, in an incident called the **Great Oxidation Event**. Although it sounds dramatic, the amount of oxygen in the atmosphere was still very low compared to today, but even so, it changed the biosphere forever. We're not quite sure whether the first cyanobacteria appeared at the same time as the Great Oxidation Event or much earlier, because oxygen couldn't build up in the atmosphere until certain other molecules had first been removed.

For oxygen to build up, a planet has to be free from elements that are keen to react with it. Iron is the most common element on Earth, and if you leave an iron nail outside in the rain, it quickly reacts with oxygen to produce iron oxide, also known as *rust*. Because there was so much iron on the early Earth, it would have reacted with the oxygen produced by the first cyanobacteria, and sure enough, huge deposits of iron oxide – called *banded iron formations* – started to form in the early oceans more than three billion years ago. Then once all the available iron atoms had been locked up in iron oxide, the levels of oxygen in the atmosphere could finally start to increase.

The oxygenated atmosphere delivered other benefits to life on Earth, one of which was the creation of the *ozone layer*. High above the Earth's surface, oxygen molecules break apart, forming short-lived molecules of *ozone* (O_3) which absorb damaging ultraviolet radiation from the sun. This natural shield prevents high-energy radiation from reaching the surface, where it would expose the long DNA strands in our genomes to significant and widespread damage. So, with oxygen widely available and genomes protected from massive solar disruption, the stage was set for the rise of multicellular beings, like animals. But they certainly took their time.

Surprisingly, the first multicellular beings did not even begin to show their faces until around 800 million years ago, so it seems reasonable to ask what the heck took them so long? One reason for the long delay is that all familiar multicellular life forms, like plants, animals and fungi are made from a third type of cell that didn't appear until around 1.8 billion years ago. These cells are called *eukaryotes* (pronounced: *you-carry-oats*) and they are thought to be strange hybrids of archaea and bacteria.

The interval between 1.8 billion and 800 million years ago defines the time between the first appearance of the eukaryotic cell and the first multicellular beings that could loosely be called animals. Rather rudely, it is dubbed the *boring billion* because scientists can't really decide why it took eukaryotic cells so long to gang up – although they aren't short of ideas.

One suggestion is that that the rise of animal life was impossible until levels of atmospheric oxygen approached the levels seen today. The amount of oxygen in the atmosphere jumped up again around 800 million years ago, just before the first animals appeared in the oceans, and many think that this can't be a coincidence. But experiments with modern sponges, widely believed to be the most primitive form of animal, show that they can grow quite happily in levels of oxygen very similar to those that prevailed around 1.8 billion years ago, so it's not clear whether lack of oxygen really held them back.

An alternative explanation is that eukaryotic cells needed one billion years to perfect their own bag of tricks. What goes on inside cells is far more complicated than the mechanisms needed to hold cells together, but cells can't gang up without some co-ordination. So, to see how cells regulate their internal affairs, let's take a closer look at those masters of the single life – the organisms that we most love to hate: bacteria.

Chapter 5

BACTERIA

The good, the bad and the ugly

Animals eat an astonishing range of foods, from plants to other animals, and humans are no exception. Many of us spend quite a lot of time planning our next meals, but once swallowed, we try to avoid thinking about what happens next. In our muscular stomachs, food is pounded around, and both here and in the small intestine, enzymes begin the process of digestion by breaking down our unfortunate victims into building blocks that we can absorb and use. But we wouldn't be able to extract all that we need without the help of those chemical masters, bacteria and archaea.

Inside the large intestine, in total darkness and a long way from the nearest oxygen supply, astonishing numbers of bacteria flourish, packed together side by side. Most are busy breaking down tough fibrous material, like *cellulose*, which plants use to reinforce their cell walls. Our bodies can't produce the enzymes to digest cellulose but there are many different types of bacteria that can, and the competition between them to gain access to this desirable material is fierce. Inside our guts, unseen and unheard, our bacterial partners wage war on each other as they battle over precious resources, releasing toxins and even stabbing their opponents with poisonous harpoons.

The gut forms a long muscular tube, and contractions of the walls cause restless movements that continually churn the bacteria around. About once per day (depending on diet) the tube dramatically opens at one end and any undigested food from the large intestine is unceremoniously ejected, along with enormous numbers of unfortunate bacteria. But, at the other end, the human host is busy topping the tube up with fresh material, so the bacterial populations quickly replace their lost numbers.

The presence of so many voracious bacteria within the gut doesn't indicate ill-health or disease. It has been estimated that a typical human body contains one bacterial cell for each human cell, and many of these are to be found in the gut. All healthy humans support enormous populations of bacteria that help them to digest difficult foodstuffs and manufacture key vitamins that human cells simply can't make for themselves. Together, this community of bacteria is called the *microbiome*, and it's causing a stir in medical circles, as people now realize just how important our fellow travellers really are.

There are small numbers of archaea in the human gut too – producing the gas methane – and a smattering of others cell types, but the vast majority of its inhabitants are bacteria, with whom we have an ongoing love-hate relationship. A very small number of bacteria are actually harmful to humans, but most are harmless and we certainly couldn't live without them. They have been the dominant life form on Earth since life began, and they will still be here when we are nothing more than a few fossilized traces. So, what can we learn from these ultimate survivors?

One type of bacteria that we understand better than most is called *E. coli* (pronounced: *Ee coal-eye*) – often found living harmlessly in the guts of humans and occasionally causing disease. Like many bacteria, *E. coli* lives fast and dies young. When things are going reasonably well, it can turn itself into two daughter cells in about the time it takes to watch an episode of your favourite sitcom. But even the writers of the most far-fetched sitcom would be amazed at what a single bacterium crams into its short life.

Before it can turn itself into two cells, *E. coli* has its work cut out. To supply two daughter cells with all the things they need, it has to copy its genome, double up the ribosomes, produce enough extra cell wall and membrane to encase two cells and manufacture around two to four million new proteins. To perform this spectacular feat, it will have to consult its genome, which in most bacteria is a streamlined affair requiring only a single circular chromosome.

As we have seen, the genome of all cells, whether bacterial or human, is made up of thousands of individual instructions, called **genes**. Each gene contains the code to build a single protein; but a cell can't survive by instructions alone. It needs co-ordination.

Like any good factory, cells need to have the right tools and machinery in place at all times. The instructions to make thousands of proteins are stored in the genome, but if all genes continually fired out messages to the ribosomes, the cell could easily descend into chaos. As media-savvy humans have recently discovered, controlling the message is key, so one important way that cells control their operations is to regulate the messages that genes send out.

To produce an RNA message, a group of enzymes must first bind to the gene.

This sounds simple enough, but the genome is just a continuous stretch of DNA, so the first problem is knowing where individual genes start and stop. The enzyme in charge of building RNA messages only knows where to attach itself because each gene begins with a special sequence of letters called a *promoter* – essentially a large, flashing sign that says 'Start here!'. If the promoter is clearly visible, then the enzyme will bind to it and produce RNA messages that are sent for translation. But if the promoter is blocked, the enzyme will simply pass by and ignore the gene entirely.

To keep the cell running smoothly, the promoters of essential genes, like those involved in respiration, are normally set to the 'on' position, with their 'Start here!' signs exposed. In contrast, genes for emergencies or special occasions are set to 'off', meaning that the proteins they encode are not normally produced. This selective blocking of emergency genes means that, most of the time, only the essential, or *housekeeping*, genes are actually in use. But this can change rapidly when something goes wrong.

Genes are Labelled with 'Start Here' Signs

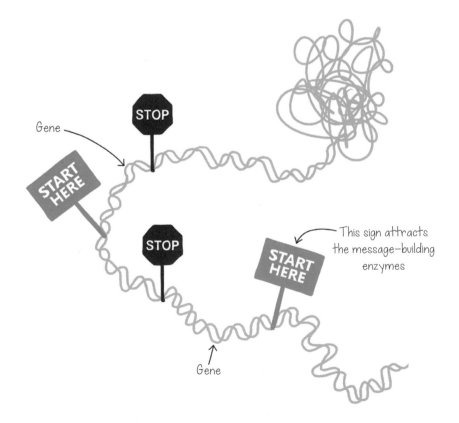

The preferred food of *E. coli* is glucose, but it can't always get what it wants. When glucose is unavailable, alternative sugars might be present, like lactose, which is commonly found in milk. With a little processing, lactose can be a useful energy source for *E. coli*, but it needs two additional enzymes to convert the lactose into glucose. Tucked away inside its genome are the instructions to build these enzymes, but until they are needed, the cell just keeps them under wraps.

The genes that code for the lactose-processing enzymes are grouped together under the control of a single promoter. Normally, the promoter is covered up by a special *repressor protein* that binds firmly to it, preventing the enzyme that builds messenger RNA from attaching itself and keeping the genes silent. The only way to remove the repressor protein is to change its shape, so that it no longer fits snugly onto the promoter, and the neat way in which this happens reveals how cells have evolved highly targeted responses to particular signals.

The shape of a protein can be changed when another molecule binds to it, by pushing and pulling the amino acids within the chain into a different configuration. In this case, the repressor protein has evolved to change its shape in response to lactose. When lactose is absent, the repressor protein clings tightly to the promoter, but when lactose appears within the cell, it binds to a different part of the repressor protein and alters its shape. With lactose firmly bound, the repressor protein is now the wrong shape to bind to the promoter, so it falls off and reveals the 'Start here!' sign. The newly revealed sign now attracts the attention of the message-building enzyme, and RNA messages are sent to the ribosomes, which then build the enzymes needed to process the incoming lactose. But, as soon as lactose disappears, the repressor protein will be the right shape to bind to the promoter and the genes will again fall silent.

The ability to switch genes on and off is known as *gene expression* and it's one important way that cells can respond to new challenges. Some bacteria, like *E. coli*, switch on (or express) different genes when a new food appears, but bacteria respond to hundreds of signals, including low oxygen levels, high temperatures and other unpleasant surprises. And a few can sense some pretty extraordinary things and make sweeping changes – with serious consequences for other inhabitants of our planet.

Lactose Regulates Gene Expression in Bacteria

Lactose absent

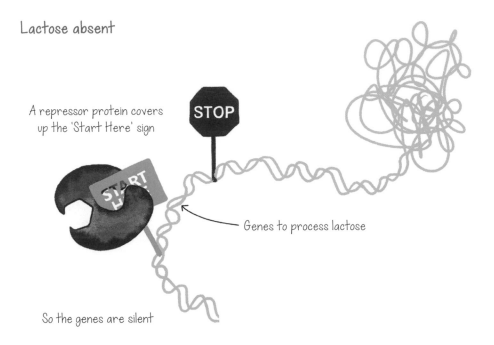

A repressor protein covers up the 'Start Here' sign

STOP

Genes to process lactose

So the genes are silent

Lactose present

Lactose binds to the repressor protein, causing it to change shape, and fall away

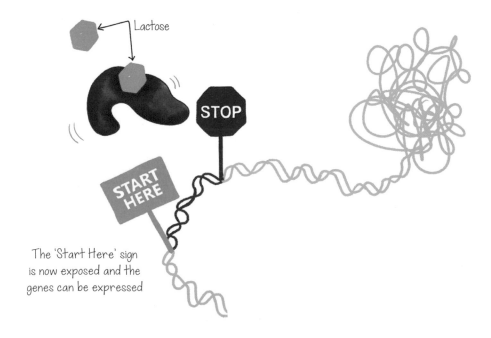

Lactose

STOP

START HERE

The 'Start Here' sign is now exposed and the genes can be expressed

Pathogenic bacteria cause disease by taking more than they should from the bodies they inhabit. Not content to live within reasonable boundaries, they secrete toxins and enzymes that damage and destroy host tissue, and so release extra food for their own growth. Only a few species are dedicated to damaging human bodies, including some famous villains, like the bacteria that cause plague or cholera. But many more are so-called *opportunists* that patrol the edges of our lives, just waiting for a chance – a chink in our armour – to change their normal behaviour and exploit any weakness. Of course, bacteria don't have it all their own way. Our bodies are shielded by a barrier of near-impenetrable skin, while cells from our *immune systems* patrol every blood vessel, waiting to pounce on intruders. If our first-line defences are breached and opportunist pathogens enter, these brave immune cells will throw themselves into the firing line and deal out death and judgement to any bacteria they find.

So, opportunist pathogens have to be stealthy. If they can sneak inside a body – especially one where the immune system isn't firing on all cylinders due to illness or old age – they will try to multiply up undetected and then launch a co-ordinated surprise attack. To succeed, they need to know when their invading army is large enough to overwhelm the host's defences. But how to do this when no one's in charge?

Amazingly, some opportunist pathogens can sense their own numbers. Each cell releases tiny amounts of a signalling molecule that effectively says: 'I'm here'. Levels of the molecule rise as the number of bacterial cells increases, so bacteria can estimate numbers by monitoring how much of this molecule is around. And when the level of the 'I'm here' molecule reaches a critical threshold, it's time for the invaders to launch their attack.

One type of opportunistic bacteria that uses this method of communication causes a lung infection called *pneumonia*. Once enough of these bacteria have established themselves, their combined shouting of 'I'm here' grows loud enough to switch on *virulence genes*, which encode molecules that maximize damage to the human body. Some are enzymes, exported from the bacteria to destroy a human protein called *elastin* – which gives lungs and blood vessels their elastic properties – while others are toxins that enter host cells to disrupt the ribosomes and cause cell death. A co-ordinated assault seems to send the immune system into a panic and gives the pathogen a much better chance of overwhelming the unfortunate host – experiments have shown that bacteria that can't sense each other are much less likely to cause serious disease – so communication is key to their success.

This co-ordination among bacteria is called *quorum sensing* and is one way that a group of mindless cells can make apparently intelligent decisions. Such abilities make bacteria truly formidable opponents and it's not just our bodies that they want to eat. Most food items have 'sell-by' dates because, once opened, bacteria will enter and start to multiply, spoiling the food and potentially making it dangerous to eat. And although the food industry is very keen to keep these bacteria at bay, their job is made harder by yet another clever response that one group of bacteria possesses.

If life gets tough, then a few bacteria prefer to sit things out and wait for better times to roll around again. This involves sensing not numbers but conditions. If food is hard to find and this continues for a few generations, then they batten down the hatches and form a long-lived *spore* that can be reactivated when times are better. Safe within this escape pod, the genome of the cell and a contingent of ribosomes lie dormant, ready to fire up again when food supplies increase.

The trigger is starvation – if growth rates are too slow then these bacteria activate several hundred genes to initiate the process of spore formation. When a spore is finally released from the mother cell, it is protected by an incredibly resistant outer wall, making it very hard to eliminate. Spores can withstand high temperatures, ordinary disinfectants, alcohol and even levels of radiation that would be high enough to kill a normal cell. Indeed, these spores are so long-lived that they have been revived from the tombs of Egyptian mummies, and claims of viable spores have been made for even older samples, dating back millions of years. However, we shouldn't be too alarmed, as most bacteria do not form long-lived spores of this kind.

So, bacteria operate a wide range of emergency responses. From fight to flight, these are enabled by batteries of emergency genes that are normally switched off, but they can easily be switched on again if the correct signal is received. This raises the question of why cells aren't packed to the rafters with emergency genes that bestow a whole range of superpowers – after all, surely there's no cost to carrying genes that are mostly silent?

An ever-expanding genome with tools for all occasions might seem highly desirable. But the genome has to be copied every time a cell divides and this takes time, energy and raw materials that cells might not be able to afford. Bacteria live in a highly competitive world – even when the good times roll, they might quickly disappear again – so it's essential to take maximum advantage of any resources that become available.

Imagine that a single bacterium has hit the jackpot and found the equivalent of a delicious bowl of ice cream. Glucose flows into the cell, which gets to work on turning itself into two cells as quickly as possible. The two daughter cells take up twice as much glucose between them as the original parent cell, so they quickly become four, and then eight. The time taken to turn one cell into two is called the *doubling time*, which can be as little as 40 minutes for *E. coli*. So, in around 13 hours (800 minutes), a population of *E. coli* can go from a single individual to just over one million cells.

Now, imagine there is a competing cell that has decided to carry a little extra baggage. This cell is packing some genes for unlikely emergencies, and because it takes longer to manufacture the extra DNA letters and to copy the expanded genome, its doubling time is five minutes longer. So, how will it fare against its sleek competitor?

The short answer is: badly. With a doubling time of 45 minutes, the cell with the expanded genome will find itself with only around 65,000 cells after 12 hours, when its competitor has just passed a quarter of a million – in other words, it is outnumbered 4:1, and as time goes on it's only going to get worse. If this outcome is repeated every time the two cells meet, then the cell with the expanded genome is facing only one possible future: extinction.

Of course, an emergency might arise for which the cell with the expanded genome is perfectly equipped to cope. If this emergency wipes out its competitor, then 65,000 cells in 12 hours looks extremely competitive compared with zero. But, the key question is: how often do such emergencies arise? If they are common enough, then we might expect the cell with the expanded genome to prosper and push the unequipped cell to extinction instead. But, if emergencies only come along rarely, then the leaner, meaner genome will win out. This is why cells are only equipped to cope with emergencies that are quite likely to happen and they are not equipped

The Cost of Carrying Extra Genes

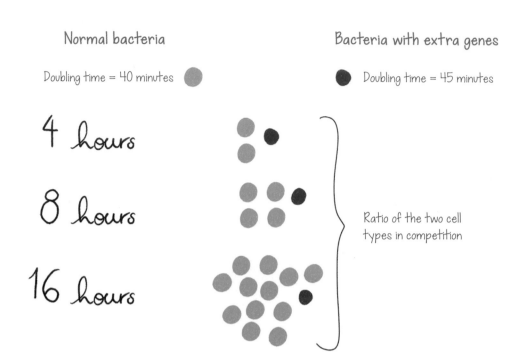

Normal bacteria

Doubling time = 40 minutes

4 hours

8 hours

16 hours

Bacteria with extra genes

Doubling time = 45 minutes

Ratio of the two cell types in competition

to cope with situations that they rarely, or never, encounter.

The ideal situation would be to customize the genome and carry extra genes only when they are likely to be needed. This is exactly the kind of flexibility that we have when we buy a mobile phone. No one would choose a phone that didn't have a few essential functions – like the ability to make phone calls, send text messages and take photographs. But, of course, people use their mobiles for lots of other things too: walkers and cyclists obsessively check the weather, foodies look up the fastest route to the nearest restaurant, and bargain hunters compare prices of everyday items. These extra functions are provided by apps, which can be downloaded and deleted without affecting the phone's core functions. And different people choose different apps, depending on whether they are cyclists, foodies, bargain hunters or whatever.

To give themselves flexible functionality, bacteria and archaea invented something similar to apps, called *plasmids*, a very long time ago. A plasmid is a small piece of DNA carrying genes for special occasions, and each bacterial cell can contain more than one. But plasmids are separate from the essential genes, which are always kept safe on the chromosome, and they can be dumped quite easily without compromising the cell's core functions — in exactly the same way as people can delete apps they no longer require without any risk of damaging their phone. One key reason that a person might choose to delete apps is if their phone is running out of storage, and this is also true for bacteria, as they want to remain as streamlined as possible. So, if the genes carried by a plasmid are useful, then bacteria that keep the plasmid will outcompete those that don't. But, if the genes carried by the plasmid are *not* useful, then bacteria that have dumped the plasmid will outcompete those that still carry it.

Plasmids have the extraordinary ability to transfer themselves between cells, allowing the special functions they carry to be spread far and wide. Indeed, bacteria will even pick up random bits of DNA from their environment, allowing them to test-drive genes that were simply left lying around by other cells. This is very unfortunate for humans as many pathogenic bacteria have picked up plasmids carrying resistance genes to drugs called *antibiotics*, which humans spent a very long time developing and that we hoped would continue to work well for decades to come.

An antibiotic is a drug that kills bacteria, and the first one was a famously accidental discovery. Alexander Fleming was growing bacteria in a lab, when he noticed that a fungus had invaded his experiments. The fungus produced a bacteria-killing chemical and about 10 years later this was refined into a drug called *penicillin* – the first antibiotic safe for human use.

The introduction of antibiotics strongly tipped the balance in the war between humans and bacteria, but today there is a very real danger that we are losing our advantage. Antibiotic resistance genes have been passed between bacteria much faster than we might ever have imagined and some plasmids confer resistance to multiple antibiotics – turning their owners into so-called *superbugs*. If a superbug colonizes a human body, then that body is in real trouble: new antibiotics just aren't being discovered fast enough, leaving people vulnerable to infections that could, until recently, have been cured.

So, how did these resistance genes spring up so quickly? Penicillin was isolated from a fungus, but several antibiotics have been isolated from bacteria. Bacteria

Inside **ruminants,** 89% of the **methane** produced by archaea is **belched** away

produce antibiotics because they have waged war against each other for millions of years, and by exporting chemical weapons into the environment, they can kill their competitors and monopolize resources. But, a bacterium that produces an antibiotic also has a real problem: it could easily poison itself. To prevent this, bacteria have developed ways to resist their own chemical weapons, including enzymes that either chop up antibiotics or alter them chemically to render them harmless – and the genes coding for these ready-made defences have been transferred via plasmids to a whole range of human pathogens – much to our concern.

One hope is that we can combat resistance by restricting the use of antibiotics. Antibiotics have been blithely handed out and far too many bacteria have been exposed to them. But if we could remove antibiotics from most environments, then more bacteria might lose their resistance plasmids – after all, we expect bacteria to lose any genes that don't confer a real advantage. To create this environment, we need to save antibiotics for real emergencies, rather than using them carelessly and to maximize profits, as happens in some types of farming.

Cattle belong to a group of mammals called *ruminants* that have developed a tight relationship with many different species of bacteria. Ruminants have been hugely successful because they can make a living out of eating a widespread type of plant – grass. Grass is full of cellulose, and as cattle don't produce the right enzymes to break it down, they have evolved a complex digestive system involving multiple stomachs to house enormous quantities of bacteria that do.

The first stomach is called the rumen (from which ruminants get their name) and this is filled with bacterial friends. Cows and other ruminants are famous for 'chewing the cud', which means regurgitating half-digested food from the rumen back into their mouths to chew it a bit more, and this is necessary because cellulose is really hard to digest and the bacteria in the rumen need all the help they can get. But in cattle feedlot farming – where animals are kept intensively in the USA and other countries – cows aren't fed grass; instead they are given pellets of artificial feed that are much easier for them to process.

When cows are fed this easy-to-digest diet, they put on weight faster, and this means higher profits. But, the bacteria in the rumen will happily consume some of this new food and use it to maintain their own populations, so if they weren't there, the cows would put on weight even faster still. This is why, in cattle feedlots, farmers give cows antibiotics, which go by the name of 'growth promoters', to suppress the bacteria in their digestive systems. Using antibiotics in this way is banned in some parts of the world, for example, in Europe, as it is considered too dangerous. The concern is that 'growth promoters' also promote the evolution of antibiotic resistance, which could then be transferred via plasmids to bacteria that are potentially dangerous to humans.

Whatever your views on the use of antibiotics in farming, doctors around the world have become more careful before prescribing them to humans. Most sore throats are not caused by bacterial infections but by viruses, which don't respond to antibiotics, so don't be surprised if your doctor is reluctant to prescribe them next time you feel a little hoarse. Indeed, because antibiotics can wreak havoc with your gut microbiome there are always costs and benefits to be weighed up, and you should always follow your doctor's advice.

Most cells carry genes for emergencies that they can turn on – or express – when needed. But switching genes on and off is a fairly crude form of regulation, and it's also rather slow. If the cell is out of balance, then a rapid response can prevent a minor setback developing into a full-blown crisis, and this is achieved in a different way – by fine-tuning the cell's *metabolism*.

Metabolism is the entire set of chemical reactions taking place within the cell. This includes reactions, like aerobic respiration, that release energy, plus all the reactions that demand energy, like building proteins or long DNA strands. Metabolism has to be closely regulated, otherwise the cell might shunt all of its glucose into aerobic respiration, for example – and so have masses of freshly recharged ATP batteries – while neglecting to redirect some of that glucose into making new building blocks for growth and repair.

To avoid the cellular equivalent of being all dressed up with no place to go, the cell makes good use of *negative feedback*, where the product (the thing being made) inhibits its own production. A familiar example of a negative feedback is a thermostat that regulates the temperature of a room: when the temperature is too high, the thermostat turns the heater off, but when the temperature is too low, the thermostat turns the heater on again. This feedback keeps the temperature in our houses more or less constant, so we don't have to continually put jumpers on and take them off again.

Cells employ negative feedback to regulate the levels of many different molecules. Too much – or too little – of just about anything causes problems for the cell, as it doesn't want to either run out of essential materials or find itself piled high with mountains of useless molecules that just get in the way.

To see how a cell regulates the flow of material through metabolic pathways, it helps to think about a factory in our world. In a modern factory dedicated to making something like a toy, there is often a production line run by robots, rather than people. Perhaps the factory is making a doll, starting with a body, which the factory imports from elsewhere. The bodies move along a conveyor belt to the first robot, which picks up the body and adds a head. The robot then releases the doll and the conveyor belt takes it along to the next robot, which screws on the left arm, before releasing it to yet another robot, which screws on the right arm. The doll can be passed along the production line until eventually the finished product – hopefully complete with all its limbs – emerges at the end.

For the production line to run smoothly, it's clearly important that each robot works at a similar speed. If it takes much longer to screw on the left arm than the right arm, then one-armed dolls will pile up between the two arm-screwing robots and pretty soon the production line will grind to a halt. One way to solve this problem is to have two robots screwing on the left arm – extra workers will inevitably make things go faster – but of course extra workers have to be built, and this costs money.

Finally, it's worth considering what might happen if there is a branch point in the production line. Perhaps dolls can go one way or the other and will be dressed in different clothes, so that the factory can produce more than one type of doll at the same time. But if the factory already has enough of one type of doll then it needs to find a way to shut down one of the branches.

In a cell, there are metabolic pathways that process molecules rather than toys. The robots are enzymes that blindly grab molecules of the right shape and make a slight alteration before releasing them again. The altered molecule is now the right shape to be grabbed by the next enzyme, which makes a further alteration, and so it continues until the end product is reached. Some metabolic pathways run in straight lines, but others include branch points, where molecules can be funnelled off for different purposes.

To prevent molecules building up and blocking the pathways, the cell makes sure that it has plenty of the main metabolic enzymes by permanently switching on the genes to manufacture these core workers. Regulating traffic through branch points is trickier and relies on negative feedbacks. A good example is the manufacture of certain amino acids.

Amino acids are essential for building new proteins. If there is a shortage of a particular amino acid, then raw materials will be funnelled off to the enzymes that make it. But, if that amino acid is plentiful, it inhibits the enzyme at the branch point by locking onto the enzyme and changing its shape so that it no longer works. This means that the amino acid can effectively regulate its own manufacture (it's as if the dolls in the human factory could control the robots that produce them) and these feedbacks ensure that molecules flow through the system in just the right way to generate all the different building blocks and energy that the cell needs – and all without anyone being in charge or giving orders.

The central metabolic pathway in many bacterial (and human) cells is aerobic respiration. Glucose is fed in at one end, and the excited hydrogen that drives the turbines to recharge ATP batteries is continually spun off. At the same time, the dismemberment of glucose creates molecules that can be turned into other building blocks, like amino acids. So how much glucose has to be processed to turn one bacterial cell into two?

The *metabolic rate* of a cell or an organism is often estimated by measuring how much oxygen they use in a fixed period of time. From this, we can calculate how many ATP batteries they used up, because three to five are recharged for every molecule of oxygen consumed. Based on its oxygen consumption, a rapidly growing bacterial

Regulating Metabolism

The cycle of reactions that breaks down glucose to release energy can also be used to make new building blocks

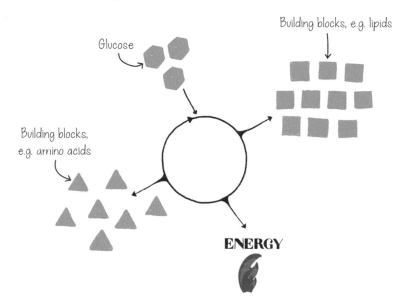

Glucose

Building blocks, e.g. lipids

Building blocks, e.g. amino acids

ENERGY

The cycle can be fine-tuned to the cell's needs

Cell needs building blocks

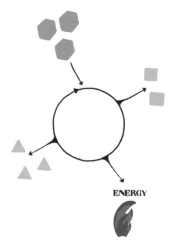

ENERGY

Glucose is diverted into making building blocks, so less energy can be generated

Cell has enough building blocks

ENERGY

Negative feedbacks prevent more building blocks being made, so more energy is generated

cell runs through roughly 10 million molecules of ATP every second – so if it takes 40 minutes for one bacterium to turn itself into two – then this doubling process will require around 24 billion molecules of ATP. And if each molecule of glucose generates 38 molecules of ATP, then it's going to take around six billion molecules of glucose to generate a new bacterial cell. So, how is it all spent?

The biggest chunk of the cell's energy budget usually goes on protein synthesis. It costs three to four molecules of ATP to add a single amino acid to a growing protein chain, and *E. coli* makes two to four million new proteins for its daughter cells with a typical length of 300 amino acids. It seems that a surprising amount of energy is also lost to leaky membranes – protons are supposed to return through the turbines that are linked to the recharging of ATP batteries – but unfortunately, quite a few of them sneak back without recharging anything. And bacteria also spend some of their hard-won energy on something that we might not expect – swimming.

There are a few Hollywood blockbusters in which humans have been shrunk to a tiny size and yet still manage to swim masterfully using breaststroke or front crawl. Well, this is truly science fiction. To a full-sized human swimmer, water is extremely well-behaved: plunge into a pool and the water cleaves easily, and once moving, a swimmer can glide for a satisfying distance before stopping. Getting going again is also straightforward – simply extend the arms and scoop water backwards, and by doing so, a decent swimmer will propel themselves forward. All too easy. But for a bacterium, things are not so straightforward.

To a bacterial cell, roughly one million times smaller than a typical human, water is considerably more challenging. Of course, water doesn't really change, but the size of the bacterium means that the forces at work are completely different. We are large enough not to notice that water is actually quite sticky – but water feels like thick treacle to our cavorting bacterium, so it will struggle to shake off the water molecules that cling to its surface.

To understand why water is sticky, we need to look more closely at the atoms within each molecule. Oxygen and hydrogen share electrons, but oxygen is bigger than hydrogen and, like big brothers and sisters everywhere, it isn't very good at sharing. Instead, the oxygen atom pulls hard on the electrons, while the two tiny hydrogen atoms struggle to keep hold of their share. This means that the shared electrons end up closer to the oxygen atom, and because electrons carry a negative charge, this gives the oxygen atom within the water molecule a slight negative charge. Similarly, the two hydrogen atoms end up with a slight positive charge, because the electrons are drawn away from their positively charged protons. The upshot of this atomic tug of war is that one end of the water molecule is positively charged and the other end is negatively charged, and in the world of electrical charges, opposites attract, causing water molecules to stick together.

The force of attraction between any two water molecules is only weak, but each water molecule is surrounded by many others, and you only have to watch an insect struggling on the surface of a pond to see that their combined pull can easily drown a small fly. So, how does our bacterium – which is thousands of times smaller than the fly – make headway?

The key to swimming at very small sizes is not to scoop water backwards but to corkscrew your way through it. Many bacteria have rigid corkscrew-like projections, called *flagella*, that rotate anticlockwise and propel them forwards. Amazingly, the bacterial flagellum is one of the few examples of a true wheel in nature – and it's a phenomenal piece of kit. One or more flagella are connected to a rotating motor embedded in the membrane, and when moving normally, the motor spins anticlockwise around 100 times per second. If the motor stops, then the bacterium will stop instantly, and if the motor then rotates clockwise, it uncouples from the flagella, which then flail around, causing the cell to tumble on the spot. With the cell now facing in a new direction, the bacterium simply fires up the motor once again in an anticlockwise direction and heads off for pastures new.

Swimming is another way that bacteria can respond quickly to their environment. Bacteria can sense many different molecules and move towards those that they want to capture and away from those that might cause trouble. This movement has been intensively studied and we know that bacteria can achieve some impressive swimming speeds. To fairly compare swimmers of different sizes, their speeds

A cyanobacteria must harvest around 10 photons to fix one molecule of CO_2

are often reported as the number of their own body lengths that they can cover each second, as clearly a bigger organism can cover more ground in absolute terms. Using this scale, *E. coli* clocks in at around 15 body lengths per second and the current bacterial record-holders can manage more like 100. Compare this with the fastest human swimmers who, despite their stylish strokes, can only manage one measly body length per second.

So, bacteria are certainly not simple. These minuscule marvels have had around 3.8 billion years to fine-tune their internal affairs and have assembled an incredible bag of tricks to deal with tough situations: squadrons of emergency genes lie dormant, ready to be switched on at a moment's notice – some even carried on disposable plasmids – and they continually monitor their environment and use this information to devastating advantage. Small changes might evoke a rapid response, such as whipping up flagella or making metabolic adjustments, but a more serious event could see the launching of escape pods in the form of long-lived spores, or a co-ordinated attack on an unfortunate host.

But have bacteria cells ever banded together to make larger structures? Or are they truly dedicated to spending their lives alone?

In truth, bacteria rarely or never live alone. Each time a cell divides, its descendants stick around, so bacteria are normally found in colonies. But within the colony, each cell remains independent – quietly continuing its life, absorbing what it needs and planning to turn itself into two cells as soon as it has stockpiled enough goodies – so the cells don't really co-operate to form a bigger structure. But there are a few bacteria that do things differently.

Some colonies of bacteria form large structures called *stromatolites*. Looking rather like large anthills, these rocky lumps form in shallow seas, and on today's Earth they can only be found in a few special places, like Shark Bay in Australia. Stromatolites are made by colonies of cyanobacteria – the inventors of photosynthesis – and because they photosynthesize, the cells need sunlight. A stromatolite begins to form when cyanobacteria grow on the surface of the sand in shallow water. Here the cells secrete mucus that binds them together, along with particles of sand and silt into slimy, sandy mats. Any bacteria that can't get enough light will die, but their descendants are mobile and can move towards the light, where they sit atop the sand grains and corpses of their ancestors. Over long periods of time, the lower layers of dead cells and sand eventually harden into rock, until mounds emerge that can be up to one metre tall with just a thin uppermost layer of living cells.

Stromatolites are truly ancient structures and could have been admired by a visiting alien who happened to drop by any time during the last 3.5 billion years. We think that photosynthesis only emerged on Earth around 3 billion years ago, so not all stromatolites may have been made by cyanobacteria, implying that living together in mats has certain advantages for more than one type of bacteria. So why not go the next step and form a true multicellular being?

To form a multicellular being, certain formidable challenges must be overcome. But perhaps the first challenge is: what exactly is the advantage to forming one in the first place? We have already seen that bacteria are phenomenally successful, so it's unclear how being multicellular could allow them to do any better.

Once again, cyanobacteria provide us with a wonderful example of how cells might be persuaded to stick together. Some cyanobacteria already live together in a long filament, consisting of many cells stuck together side-by-side. So far, so unimpressive, but what's really unusual is that not all of the cells are the same – roughly one in ten cells within the filament develop a thick cell wall and perform a completely different task from the other nine. Given that all multicellular beings contain different types of cell that carry out specialized tasks, we clearly need to know why this specialization arose and what benefits it brings.

Cells require carbon to construct basic building blocks, like carbohydrates. But other building blocks, such as amino acids, require extra elements, like nitrogen. Nitrogen is by far the most abundant gas in our atmosphere, but getting hold of useable nitrogen is an awful lot harder than we might expect.

Nitrogen makes up 80% of our atmosphere, so it really shouldn't be in short supply. But nitrogen atoms go around clamped together in pairs, and they are extremely hard to separate. Early in the history of life, cells developed the machinery to make nitrogen more available, a process called *fixing*, by breaking the two nitrogen atoms apart and combining each one with hydrogen to form ammonia – NH_3 – which can be used to form amino acids. This reaction is highly energy demanding and can only take place in the absence of oxygen – not a problem on the early Earth, where there wasn't any oxygen in the atmosphere – but for cyanobacteria that carry out photosynthesis and generate oxygen all day long, this is a very big problem indeed.

Some unicellular cyanobacteria solve the problem by only fixing nitrogen at night, when the lack of sunlight shuts down photosynthesis and oxygen is no longer released. But this probably involves some loss of efficiency – the cell might run out of useable nitrogen during the day when it has plenty of carbon, or indeed, it might run out of useable carbon during the night, when it has plenty of nitrogen. A much better strategy would surely be to form a group and divide up the tasks.

In the multicellular cyanobacteria this is exactly what has happened. Nine out of ten cells in the chain photosynthesize and duly produce oxygen as a waste product, while one in 10 cells forms a specialized structure with a thick wall that keeps oxygen out. Within these cells, nitrogen fixation can take place 24/7 without being affected by the oxygen-generating activities of their neighbours on either side. But crucially, the cells exchange riches: those that photosynthesize transfer carbon to the nitrogen-fixing cell, while it transfers a nitrogen-rich compound to its photosynthesizing buddies in return. And this kind of division of labour is crucial to the success of multicellular beings.

So, bacteria *have* made the step to multicellular beings – even if those beings are simple chains of cells. The chains begin as a single cell, but when it divides, the daughter cells stick together, rather than separating and leading independent lives. All of the cells in a chain are descended from the first cell, so they share an identical genome, but because some cells are photosynthesizing and some are fixing nitrogen, they clearly aren't all expressing the same genes.

Regulating gene expression is one of the key technical challenges that faces multicellular beings. Be they cyanobacteria, plant, animal or fungus – all multicellular beings are built from many different cell types that carry out specialised functions, and so each cell type must express different combinations of genes. But, while bacterial

A Cyanobacterial Strand

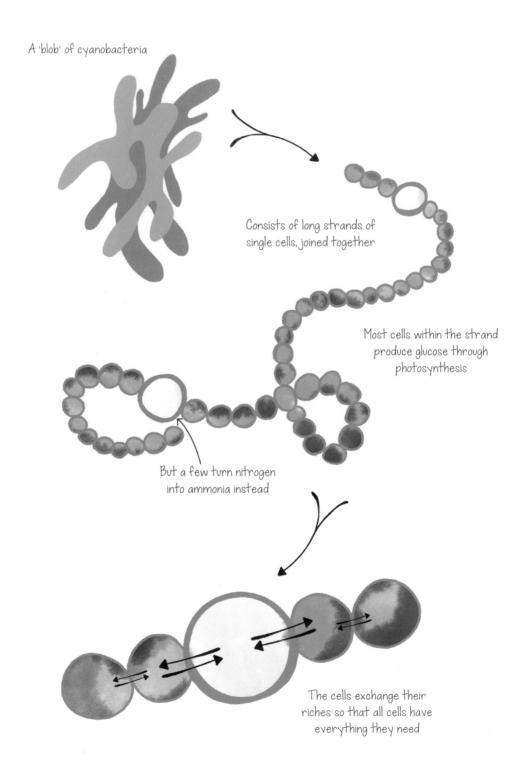

A 'blob' of cyanobacteria

Consists of long strands of single cells, joined together

Most cells within the strand produce glucose through photosynthesis

But a few turn nitrogen into ammonia instead

The cells exchange their riches so that all cells have everything they need

beings contain only a small number of cell types, our bodies contain hundreds – from long thin nerve cells that run the length of a leg, to the tiny doughnut-shaped red blood cells that carry oxygen around.

To construct a functional multicellular being made from hundreds of cell types, those cells need to be flexible in shape and form, and not just in chemistry. This probably explains why bacteria never managed to construct an octopus or a daisy, because their rigid cell walls don't allow them much flexibility, and they are stuck with simple rigid shapes, like rods and spheres. But, by solving technical problems, like gene expression and aerobic respiration, bacteria helped to give rise to the eukaryotes – a fundamentally different kind of cell – that would combine bacterial brilliance with the ability to form an extraordinary range of shapes and forms. The eukaryotes are the master builders and it's time to meet our monstrous ancestors.

Chapter 6

EUKARYOTES

We are all descended from monsters

Crawling across a surface in relentless pursuit of its undersized prey, a giant blob going by the name of *amoeba* (pronounced: *am-ee-bah*) oozes its way to victory like some monstrous slug. The bacterial victim senses the danger and desperately changes direction, but unlike a common-or-garden slug, the amoeba is hyper-flexible and continues the hunt with deadly intent. It sounds like a terrible Hollywood movie, but instead of watching a screen, we are looking down a microscope as one type of cell pursues another – and there's a horrible inevitability to this mismatched encounter. After just a few seconds the predatory cell extends voluptuous engulfing arms and encloses its bacterial prey in a deadly embrace. Closing our eyes in horror, we can almost hear the belch of satisfaction.

The pursuing cell is a eukaryote, and in every sense, it is a monster. Until this unlikely upstart made its first appearance around 1.8 billion years ago, the Earth was home to only two cell types: bacteria and archaea, and these are generally small cells with simple shapes and little internal structure. Collectively known as *prokaryotes* (pronounced: *pro-carry-oats*) they make their living by deploying creative chemistry and do not appear to possess the capacity (or perhaps the inclination) to engulf one another. Nevertheless, between them, this unlikely duo gave rise to the eukaryotes, who proceeded to rewrite the manual of what it means to be a cell.

Even by eukaryotic standards, the amoeba is enormous – around 250 times longer than the familiar bacterium *E. coli*. The difference in size can be better appreciated if we supersize the amoeba to roughly the size of a three-storey house, in which case an engulfed *E. coli* would be about the size of a bulb of garlic. Of course, the amoeba is a true giant, and a typical eukaryotic cell is perhaps only 10 or 20 times longer than a single *E. coli*, so if the typical eukaryotic cell was the size of a kitchen, *E. coli* would be about the size of a pumpkin. But, whichever comparison we make, eukaryotic cells dwarf their prokaryotic ancestors.

Enormous size has consequences for any organism. First, the internal volume of eukaryotic cells is far too large to form one simple 'bag' so they are divided up into a series of compartments using internal membranes (rather like the rooms in a house). Second, most eukaryotes have ditched the rigid cell wall and are supported instead by a dynamic internal *cytoskeleton* that reinforces the fragile cell membrane

135

and allows them to crawl, move and stretch themselves into an extraordinary variety of shapes. Third, they are fuelled by a network of internal power stations dedicated to recharging the billions of ATP batteries that they blast through every second. And finally, this type of cell is extraordinarily well-suited to building larger, more complicated beings – including every plant, animal and fungus on our planet.

The evolution of eukaryotes is one of those events in the history of life that can truly be described as momentous. Eukaryotic cells are awash with unique features that allow them to form the countless multicellular creatures that now appear to dominate the Earth. But, of course, they didn't evolve with any intention of forming multicellular beings – that was simply a brilliant afterthought. They evolved because single eukaryotic cells have incredible superpowers and can hold their own alongside both bacterial and archaeal competitors.

Eukaryotic cells are highly diverse in form, shape and behaviour, but their fundamental chemistry is remarkably similar. Mostly reliant on aerobic respiration – eukaryotes pursue energetic lifestyles by exploiting the oxygen-rich atmosphere that developed on Earth thanks to cyanobacteria. But eukaryotes did not invent aerobic respiration, they stole it, and we know precisely who they stole it from because the inventors remain trapped inside nearly every eukaryotic cell, enslaved.

Unlike most bacteria, eukaryotic cells are clearly visible under a normal light microscope, where they have a rather grainy appearance. In the middle of each cell, the genome is stowed away inside a darker, roundish blob called the nucleus, but other structures have been revealed using the *electron microscope* – a powerful tool with much higher magnification. Pictures taken using the electron microscope show that, beyond the nucleus, eukaryotic cells are not just bags of liquid where molecules bump into each other at random (which is a reasonable description of the inside of a bacterium). Instead, they are highly organized, packed with membrane-bound compartments of different shapes and sizes collectively known as *organelles*.

Prominent among the organelles are a host of small cylindrical blobs called *mitochondria* (pronounced: *might-oh-con-dree-ah*). These are the powerhouses of eukaryotic cells and each is roughly the size and shape of a single bacterium. Possessing both a small genome and their own ribosomes, their key feature is a deeply folded internal membrane studded with turbines and proton pumps. All day long, mitochondria pump protons across their membranes using energy released from the destruction of glucose and so recharge millions of ATP batteries. Once released into the cell, the batteries fuel all of its activities, so the presence of mitochondria frees the outer membrane from the task of energy generation, in sharp contrast to the cell's prokaryotic ancestors. But the nature of this seismic shift has always troubled biologists.

In 1966, biological scientist Lynn Margulis caused a stir by putting forward a radical and sinister idea about the origins of mitochondria. She proposed that these tiny cylinders were originally free-living bacteria that had been engulfed and enslaved by their monstrous host, which neatly explained both why they were exactly the same size and shape as a bacterium and why they contained their own genome and ribosomes. At first, this theory was treated with contempt – it just seemed ridiculous that one cell could enslave another; however, few scientists now seriously doubt this explanation. But if mitochondria are descended from free-living bacteria, what kind of cell engulfed them?

Our current best guess is that the engulfing cell was a type of archaea. Since the 1970s, scientists have classified all cells – and hence all life – into three major groups or *domains*: Bacteria, Archaea and Eukaryotes. More recently, genome sequencing has allowed us to reliably trace relationships among species (revealing, for example, that hippos are closely related to whales) and this technique has shown that eukaryotes share more in common with archaea than with bacteria. But this still leaves us with an amoeba-sized problem. Archaea cannot change their shape and aren't known to prey on other cells or engulf them, so how could an ancestral archaeon have gobbled up a bacterium 1.8 billion years ago?

Prior to the disturbing revelation that they could be *our* ancestors, most people knew rather little about archaea – and probably cared even less. But scientists were soon scurrying around in the hope of raising the status of archaea from obscure weirdos living in extraordinary places to mainstream actors, preferably with godlike

The Key Differences Between Bacteria and Eukaryotic Cells

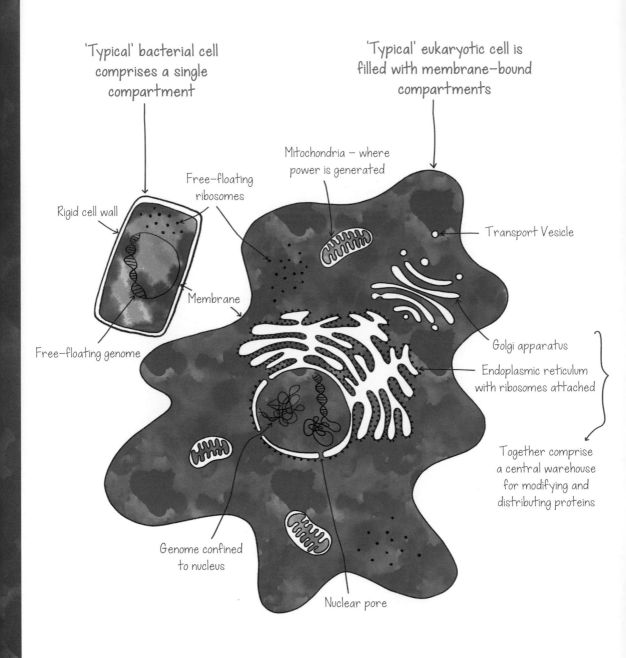

'Typical' bacterial cell comprises a single compartment

'Typical' eukaryotic cell is filled with membrane-bound compartments

Mitochondria – where power is generated

Free-floating ribosomes

Rigid cell wall

Transport Vesicle

Membrane

Golgi apparatus

Free-floating genome

Endoplasmic reticulum with ribosomes attached

Together comprise a central warehouse for modifying and distributing proteins

Genome confined to nucleus

Nuclear pore

qualities. In the rush to find and classify archaea, scientists analysed samples from all over the world, hoping to find a likely ancestor for the eukaryotes. Given the affinity of archaea for strange places, one such location was a system of deep-sea hydrothermal vents called *Loki's Castle* in the mid-Atlantic Ocean. Loki is a Norse god, whose modern popularity owes a lot to Marvel comics, and despite his reputation for making mischief, his deep-sea home didn't disappoint the cell hunters.

Loki's Castle was sheltering a group of archaea that were previously unknown to science. They were named the *Asgard Archaea* – after the mythical home of the Norse gods – and they possess certain features that were previously thought to be unique to eukaryotes. Even more amazingly, in 2020, a Japanese team of scientists released photographs that shocked the scientific world. They had spent 12 years culturing Asgard Archaea from the deep sea, and their images revealed that instead of having simple geometric shapes – like every other bacterial or archaeal cell – they possessed extendable spaghetti-like arms. This was the smoking gun. If some types of archaeal cells have the potential to reach out and engulf other cells, then this vindicates Lynn Margulis, and puts beyond doubt her idea that eukaryotes began life as archaeal cells that captured bacteria and put them to work. But like all creative partnerships, they became much more than the sum of their parts.

In Norse mythology Loki himself is a shape-shifter, a quality fully embraced by eukaryotic cells. As we have seen, eukaryotic cells have mainly ditched their rigid cell wall in favour of a flexible internal cytoskeleton, a dynamic spiderweb of filaments that stop the cell collapsing and allow it to change shape and crawl around – and even engulf bacterial prey, as modelled by the predatory amoeba.

A free-living amoeba is easy to find in soil or in the mud at the bottom of ponds. But many other eukaryotic cells have adopted an amoeba-like lifestyle and make their living by crawling around, engulfing smaller cells – some of them in unlikely places. The immune system of humans includes specialist bacteria-hunting cells, called *macrophages*, which crawl around the body, squeezing in and out of blood vessels on the look out for invaders, and they too can extend 'arms' and engulf anything they don't much like the look of. But how on Earth can they squeeze themselves into such an extraordinary array of shapes?

Amoebas and macrophages crawl by extending so-called 'false feet' or *pseudopodia* (pronounced: *soo-doh-poh-dee-ah*). These are simply blobby extensions that extend out from the main body of the cell and adhere to the surface, allowing the cell to pull itself forwards. The cell can extend pseudopodia in any direction by rapidly assembling tiny cytoskeleton filaments that push against the membrane as they grow, causing it to bulge outwards. But, like a clumsy Fred Astaire, they can easily retract one foot and put out another, because the filaments are highly dynamic and can spontaneously assemble and fall apart again, even when isolated from a cell and kept in a test tube.

The filaments of the cytoskeleton allow other forms of movement in addition to the slow crawling of the amoeba. Many eukaryotic cells possess one or more long-lived filaments that protrude from the outer membrane, forming long flagella. The name flagellum derives from the Latin word for whip, which makes sense for eukaryotic cells, because their flagella flail back and forth in a whip-like fashion, unlike the rigid corkscrew-like flagella possessed by bacteria (which probably should have a different name). Some eukaryotic cells have a few long flagella, but others have covered themselves in hundreds of shortened versions, called *cilia*. These beat in co-ordinated waves sending cells like the predatory *Paramecium* (pronounced: *Para-me-see-um*) spiralling forwards in pursuit of prey.

A drop of pond water viewed under an ordinary light microscope will often reveal hundreds of ciliated cells zipping around in all directions. Although the magnification isn't high enough for us to see their bacterial victims, we already know that by twirling its rigid helical flagellum, *E. coli* can manage a speed of around 15 body lengths per second. So, how does *Paramecium* fare in our cellular swimming championships? The *Paramecium* might be embarrassed to know that its cilia only allow it to swim at a speed of around five body lengths per second, but unfortunately for its bacterial prey, the much greater size of eukaryotic cells means that they still cover more ground and have no trouble chasing them down.

So, the cytoskeleton is essential to eukaryotic cells: providing internal support and enabling movement – from amoeba-like crawling to impressive spiral swimming.

Movement in Eukaryotic Cells

Pseudopodia

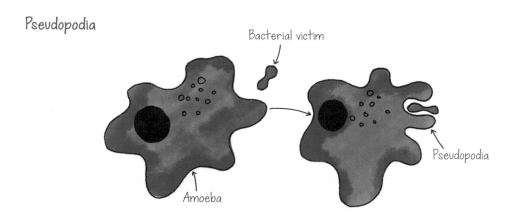

Bacterial victim

Pseudopodia

Amoeba

The amoeba can sense a potential prey item and extend pseudopodia, or false feet, to move towards it.

Cilia

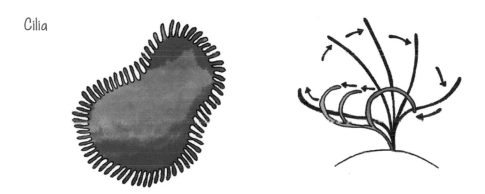

Paramecium is covered in tiny hair-like structures, called cilia, that beat in co-ordinated waves.

Flagella

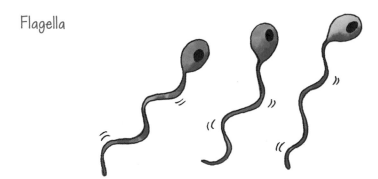

A sperm cell has a single whip-like flagellum that propels it forwards.

But, amazingly, the role of the cytoskeleton doesn't stop there: it shunts things around *inside* the cell too, forming a system of guided rails for the targeted delivery of large items of cargo.

In comparison to the scale we live at, all cells seem tiny, but the difference in size between *E. coli* and the amoeba is the difference between a bulb of garlic and a three-storey house. Size differences of this magnitude have profound impacts on nearly every aspect of a cell's biology, forcing eukaryotic cells to entirely restructure their interiors. Inside cells, molecules are continually in motion, which is essential to bring important parts together (like ribosomes and RNA messages or enzymes and their substrates). Better, a physical process called *diffusion* ensures that molecules flow from places where they are plentiful to places where they are scarce. Prokaryotic cells can rely on diffusion alone to deliver molecules to where they are needed, but scale a cell up from the equivalent of a garlic bulb to a three-storey house, and diffusion alone simply isn't going to cut it.

To understand how diffusion works, let's begin with a simple thought experiment. Humans, like all animals, are built from eukaryotic cells that carry out aerobic respiration. But, aerobic respiration can't take place without oxygen, which thanks to the inventiveness of cyanobacteria, isn't hard to find. A person sitting idly in the corner of a room will draw in lungfuls of air and some of the oxygen that they breathe in will be used by the cells in their body, so the air they breathe *out* will contain less oxygen than the air they breathed *in*. So, it seems logical that if they stay in the same place, the air around them will become oxygen-depleted. But there are no public information films warning people of the dangers of suffocation if they sit still for too long. Yes, we are advised to be active, but not because we might otherwise run out of oxygen – why not?

There are two reasons why depleting the oxygen in a room is more or less impossible. First, any room in a normal house or school is not completely sealed and so connected to the atmosphere: a vast pool of air containing an almost limitless supply of oxygen. Next, the second law of thermodynamics demands a maximum state of disorder, so the oxygen molecules within that vast pool of air will distribute themselves evenly and not cluster in a few places or leave gaping holes in others. So, as the number of oxygen molecules in the air is reduced, others move in from the wider atmosphere to replace them, and a person can sit quietly all day long (with a large packet of crisps if necessary) and not worry about suffocation – although there are other good reasons why they probably shouldn't do this!

Diffusion is probably one of the most important – if underappreciated – processes on the planet. The best thing about diffusion is that it's free: if respiring cells use up oxygen, then more will arrive with absolutely no effort on their part. Similarly, the carbon dioxide they produce will simply diffuse away, taking their waste problem with it. But diffusion has limits, and they are pretty sharply defined.

Cells are usually surrounded by water rather than air. Oxygen is soluble in water, but a bucketful of water contains much less oxygen than a bucketful of air. If oxygen is depleted by *Paramecia* whizzing around in a pond, then diffusion will replace it – BUT (and this is a big but) – it typically takes 10,000 times longer for a molecule to diffuse the same distance through water as it does in air. So, *Paramecia* really do need public information films about the dangers of sitting still. It's perfectly possible for them to suffocate if they end up at the bottom of a pond where the activities of billions of respiring bacteria have effectively depleted the oxygen. In time, diffusion will come to their rescue, but that can take a while, and it's probably one reason why *Paramecia* have cilia – not to pursue prey but to escape from oxygen-depleted water.

Diffusion is much slower in water than in air because water is a liquid rather than a gas. In a liquid, molecules are tightly packed together, but in air they are widely spaced. An oxygen molecule finds it far easier to move through air because it doesn't collide so often with other molecules, while in water it is continually buffeted around. These collisions slow molecules down, and the larger the molecule, the longer diffusion takes. So, while cells can rely on diffusion to provide a wonderful free delivery service for small molecules over short distances, it only goes so far.

The time taken for a molecule to move by diffusion also depends on the distances involved. Let's imagine that it typically takes 1 second for a molecule to move just 1 millimetre. We might therefore expect the same molecule to move 2 millimetres in 2 seconds, 4 millimetres in 4 seconds, 8 millimetres in 8 seconds, and so on; but, unfortunately, this is not how things work. Instead, the time required for a typical molecule to move by diffusion rises by the *square* of the distance: so, it will actually take 4 seconds to move 2 millimetres, 9 seconds to move 3 millimetres and 16 seconds to move 4. So, now we come to the nub of the problem. Can diffusion move molecules far enough and fast enough to provide a reliable delivery service within cells?

Let's assume that eukaryotic cells are typically only 10 to 20 times the length of a bacterium (although the comparison between *E. coli* and the amoeba makes it clear that the size difference can be much greater). But even these lower estimates translate into a very big problem for eukaryotic cells. If they rely on diffusion alone (as prokaryotic cells do), then the time taken to move a molecule from one end of the cell to the other will be one hundred to four hundred times greater, so a journey that would take one second in a bacterium will now take a few minutes – and that's just not fast enough for busy cells with active lifestyles.

How the Time for Diffusion Scales with Distance

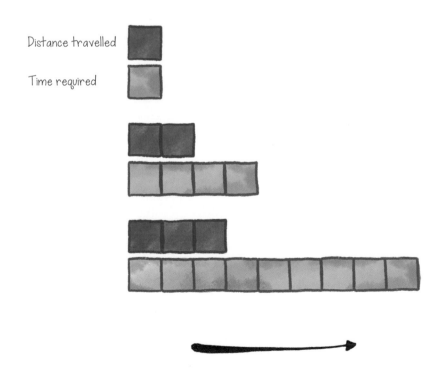

Distance travelled

Time required

To speed things up, eukaryotic cells actively move some items using filaments of the cytoskeleton as railway lines, particularly large items that will only diffuse very slowly indeed. The downside to this long-distance transport system is that, unlike diffusion, it doesn't come for free. Instead, ATP batteries must be spent on driving molecular motors that ratchet up and down the filaments, holding the precious cargo above their heads. But, on the plus side, the cargo can be targeted to a precise destination and is guaranteed not to get trapped in the wrong place. The cytoskeleton is dynamic, so these railway lines can be quickly ripped up and relaid at a moment's notice, allowing eukaryotic cells to quickly remodel their interiors, should their needs suddenly change. And one of the commonest items to be transported along the filaments are various small organelles.

A eukaryotic cell consists of a central nucleus and the surrounding *cytoplasm*. Within the cytoplasm, around half the space is taken up by a watery liquid called the *cytosol*, while the rest is filled with a labyrinthine arrangement of interconnected compartments fashioned from membranes, called organelles. Organelles are unique to eukaryotes – bacteria manage just fine without them – and they are yet another adaptation that allow eukaryotic cells to function smoothly at enormous sizes.

There are around ten different types of organelles in eukaryotic cells. Each has a specialized role to play – from energy generation to digesting the cell's victims – and it's vital that they receive the correct proteins to perform their proper functions. One way that eukaryotic cells ensure that the right proteins arrive at the right destination is to deliver them in specially marked transport vesicles. These tiny membrane-bound bubbles carry information on their membranes telling them exactly where to dock, but how do the right proteins end up inside the right transport vesicles?

Taking up around 15% of the total cell volume is an elaborate warehouse for finishing, organizing and distributing proteins to the correct locations. The warehouse has two interconnected parts whose technical names are the *endoplasmic reticulum* and the *Golgi apparatus* (pronounced: *Gol-jee*), and between them, these two extensive organelles make up around 60% of the total area of membrane within a eukaryotic

Inside eukaryotic cells, bacterial slaves keep the machinery whirring

cell. The endoplasmic reticulum receives the proteins directly from the ribosomes and they are then passed on to the Golgi apparatus for further modifications and onward transport to their final destinations.

Whether or not a protein enters the warehouse is determined early in its synthesis. As the ribosome joins amino acids together, a protein chain emerges, and if this contains the right sequence of amino acids, it will lock onto the warehouse membrane and slip inside via a special pore. Once inside, the finishing can begin. In around half of the proteins, sugar molecules are attached, which can be used as recognition tags. But the main function of the warehouse is to help the protein fold into its proper final shape. Help is needed because eukaryotic cells produce larger proteins than prokaryotes, and a longer chain of amino acids struggles to spontaneously fold itself properly, but inside the warehouse there are *chaperone proteins* to reduce the number of mistakes.

Unfortunately, despite the best efforts of the chaperones, up to 80% of some eukaryotic proteins fail to fold properly and end up malformed. Malformed proteins could rampage around the cell causing damage or obstructing normal operations, but because they are confined within the central warehouse, they can be marked for destruction and their amino acids parts recycled. Indeed, sometimes this process works almost too well.

Sufferers of cystic fibrosis have inherited a mutation in the gene that codes for the CFTR protein. In non-sufferers, this huge protein forms a channel through the outer membrane of mucus-secreting cells – like those that line the air passages of the lungs. During its synthesis, the CFTR protein is sent for finishing to the warehouse, but in sufferers of cystic fibrosis, the protein has a missing amino acid and no matter how hard the chaperones try, it stubbornly refuses to take on an acceptable shape. It is therefore endlessly rejected and recycled, but scientists believe that the 'faulty' protein would actually work perfectly well, if only the quality control standards in the warehouse were just a little lower.

Some proteins that cells produce, like digestive enzymes, could cause significant damage if they ended up in the wrong place, and the distribution system can prevent such disasters from happening. To see how this works, let's return to our amoeba as it catches up with its bacterial prey. After the pseudopodia reach out and engulf them, the bacteria find themselves imprisoned within a membrane-bound vesicle, where – like goldfish in a bowl – they await the amoeba's next move. Of course, the amoeba could release its bacterial captives into the cytosol and deal with them there, but no eukaryotic cell wants bacteria wandering around, helping themselves to the cell's contents. Instead, the amoeba prepares to finish off the bacteria within their bubble-prison, and this means bringing in digestive enzymes to reduce them to a handful of molecules.

Back in the warehouse, the digestive enzymes have passed quality control and are loaded into a small transport vesicle. To ensure that it docks and unloads its cargo in the right place, the transport vesicle is primed to recognize signature proteins displayed on the vesicle containing the captured bacteria. On arrival, the two vesicles fuse and the enzymes quickly get to work on breaking down the large molecules from which the bacteria are made, releasing building blocks that the cell can eventually reuse. But if these enzymes escaped from the vesicle, there is a real danger that the cell might digest itself.

Confining powerful enzymes to safe spaces is one of the great advantages of organelles. In theory, this confinement should prevent accidental self-digestion, but it's always possible that the enzymes could wriggle free. To prevent them wreaking havoc, the digestive enzymes don't work properly outside the vesicles, so if they end up anywhere else, they can't do too much damage. This might seem like magic, but enzymes only work well when the conditions are just right, and they are sensitive to changes in temperature and *pH* – which is a measure of acidity. The amoeba's digestive enzymes need acid conditions to work properly, so the cell activates the enzymes by adding acid to the vesicles containing food, while the rest of the cell remains neutral – and if the enzymes escape, they simply won't work very well.

Having safely docked and delivered its contents, the now empty transport vesicle leaves the bacteria to their fate and returns to the main protein sorting centre. To keep up with the supply of new vesicles, the warehouse synthesizes nearly all of the lipids that the cell needs, including those required to make new cell membranes, and amazingly this process goes on within the membrane walls of the warehouse itself.

Given that bacteria are confined within the cell prior to digestion, we can

How an Amoeba Digests its Bacterial Prey

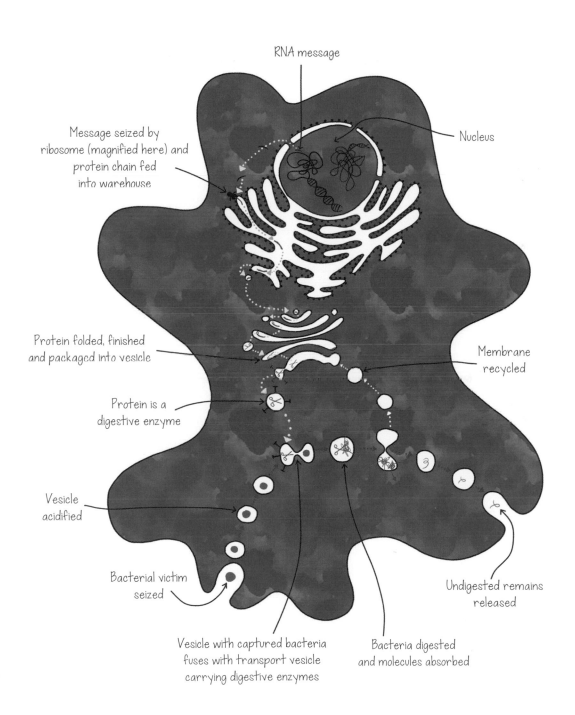

RNA message

Nucleus

Message seized by ribosome (magnified here) and protein chain fed into warehouse

Protein folded, finished and packaged into vesicle

Membrane recycled

Protein is a digestive enzyme

Vesicle acidified

Undigested remains released

Bacterial victim seized

Vesicle with captured bacteria fuses with transport vesicle carrying digestive enzymes

Bacteria digested and molecules absorbed

perhaps see how one type of bacteria ended up with a permanent home inside the earliest eukaryote, and eventually became the mitochondria. If an early eukaryote had engulfed bacteria that performed a useful service – like recharging ATP batteries – then perhaps delaying their digestion would have given the cell an advantage. Over time, this delay could have got longer and longer, until they didn't get digested at all. And there's one last piece of evidence to support this idea: unlike other organelles, mitochondria are still surrounded by an outer 'extra' membrane, a remnant of the bubble-prison in which they were originally captured.

Not all proteins are sent through the central warehouse: some – like the proteins destined for the mitochondria – are made entirely within the cytosol and delivered by diffusion, which might reflect their strange origins as captured bacteria. But, whatever the reason, by channelling proteins through different pathways, eukaryotes have streamlined their operations and ensured that the right proteins are delivered to the right destinations.

Looming over this elaborate factory is the inevitable genome, shut away inside the nucleus and acting as a head office. As genetic sequencing has become cheaper, it has become clear that it's not just the interiors of eukaryotic cells that are so different to their prokaryotic counterparts. The eukaryotic genome is a world away from its bacterial and archaeal ancestors, although precisely how these differences evolved is often hard to fathom and still the subject of intense scientific research.

We have already seen that bacteria are under tremendous pressure to keep their genomes streamlined. In support of this idea, their instruction manuals usually fit comfortably onto a single chromosome, while the genomes of eukaryotes are much larger and usually spread across multiple chromosomes (a single copy of the chimpanzee genome requires 24). Given their increased complexity, it seems perfectly reasonable that eukaryotes would have larger genomes than bacteria, but it turns out that the size of the genome doesn't reveal too much about the sophistication of its owner.

Genome size has now been estimated for thousands of species. To compare them fairly, we use the number of letters required to encode a single copy of the manual (although most cells in sexual species have a double genome that consists of two copies). It's certainly true that bacteria and archaea have relatively compact genomes – coming in anywhere between 1.5 and 5 million letters (for perspective, the Bible has around 3 million letters). In comparison, the genome of baker's yeast – a single-celled eukaryote that can be found in any kitchen – needs around 12 million letters to be written out in full. So, what happens to genome size when we move from a single-celled eukaryote to the multicellular creatures that those inventive cells constructed?

Scientists know a lot more about some organisms than others. There are two multicellular creatures that zoologists have studied intensively: a type of worm called a *nematode* and a fruit fly that you may have seen hanging around your bananas. The genome of the nematode worm clocks in at around 100 million letters, while the fruit-fly genome edges ahead with 140 million. Plant scientists have their favourites too, and the genome of a small plant called thale cress ties with the fruit fly on 140 million, while a commonly studied moss has more like 500 million. Moving on to vertebrates, mice have been unfortunate enough to feature in a huge volume of medical research, and they leave the fruit fly and the moss on the starting blocks with a massive 2.8 *billion* letters. And at the top of this pyramid of well-studied organisms stands *Homo sapiens* (the scientific name for humans) with a gigantic 3.2 billion letters (about 10 copies of *Encyclopaedia Britannica*). So, it does indeed seem that the more complex and sophisticated the organism, the bigger the genome. But get ready for a nasty surprise.

In 2010, the scientific world was rocked when a group of researchers declared that a single copy of the marbled lungfish genome contained around 133 billion letters – 38 times more than our own. And it's not just other animals that are vying for the largest genome crown – a small Japanese plant has a genome 50 times larger than ours, and many other plants, animals and fungi similarly outdo us. This seems bizarre, as it's hard to understand why building a marbled lungfish requires 38 times more information than building a human. But if a genome can be compared to a book, it turns out to be a mighty strange read.

Imagine we had a genome laid out in front of us. We might expect the genes to be found one after another, with clear start and finish points to mark where they begin and end. Between the genes, it seems reasonable for there to be some 'filler' – perhaps a few lines of nonsense just to separate them out – and this is pretty much what bacterial genomes look like, as they have been honed for maximum efficiency. But the genomes of eukaryotes are quite different.

If we tried to read a eukaryotic genome, we might find our first gene on page 1, but the next one might be pages and pages away. Between the genes we would find reams and reams of repetitive letters that don't seem to contain any useful information. Worse, even *within* the genes there are usually several sections of this stuff, which have to be removed from the messenger RNA before it can be sent for translation. Indeed, reading the eukaryotic genome is like reading a book written partly by a writer and partly by a naughty cat that likes to walk over the writer's keyboard every time their back is turned. And it seems that no editor ever bothered to remove the cat's contributions.

The general term for DNA that does not code for proteins is *non-coding DNA*, although the term *junk DNA* is also applied. When scientists first began to investigate genomes, no one really knew what this was or whether it had any useful role, but many organisms certainly have enormous quantities of it – and some more than others. Indeed, the difference in genome size between most eukaryotes isn't due to differences in the numbers of genes, it's due to differences in the size of the non-coding regions. In other words, a chimpanzee's genome is 50% larger than a dog's, but this isn't due to a difference in the efforts of the writer – they both have around 19,000 genes – it's due to differences in the contributions of the naughty cat. And humans can't redeem themselves by boasting the most efficient eukaryotic genome (with the smallest proportion of non-coding DNA) as that accolade currently goes (for reasons understood by nobody) to a pufferfish.

The extent of non-coding DNA is truly amazing – around 98% of the human genome is non-coding, although some single-celled eukaryotes, like yeast, have much less. Yeast cells, like bacteria, have been selected for fast generation times, where a streamlined genome is crucial. But most multicellular creatures are not under extreme pressure to divide as quickly as possible, so it seems that this has allowed the extra DNA to build up and gradually accumulate.

The origins of non-coding DNA vary among organisms. Within the genomes of most eukaryotes are the remains of ancient viruses that integrated themselves into the host's genome long ago, but then suffered mutations to render them harmless. But in humans, most of the non-coding DNA is due to *selfish genetic elements*. Discovered by Barbara McClintock, these are sequences of letters that simply copy and then reinsert themselves somewhere else in the genome, and so multiply up without producing anything useful for the cell. This may seem bizarre, but natural selection will favour anything that is successful – and a gene that can copy itself onto every chromosome will undoubtedly spread through a population rapidly, as long as it avoids inserting itself into the middle of essential genes too often (and thus potentially killing its host). It's also easy to see how they could spread more easily as non-coding DNA builds up. As non-coding DNA becomes an ever-greater part of the genome, then new copies of the selfish elements are ever less likely to end up in essential genes, and if they don't cause any real damage, natural selection won't remove them, allowing them to reach today's preposterous levels.

While all non-coding DNA used to be labelled 'junk', this isn't strictly true – perhaps because cells simply got used to it being there and press-ganged at least some of it into service. Some non-coding DNA has been found to play a role in regulating gene expression. Eukaryotic cells don't switch on all their genes at the same time: many are only activated when conditions are right. But eukaryotic genomes are so full of nonsense that finding the right gene may require a little help. Just like their bacterial ancestors, eukaryotic genes have promoters – the 'start here!' signs that attract the enzyme that builds the RNA messages – but because the genome is so large, it's rather like finding a needle in a haystack, and much of the DNA is wound up so tightly that it can be difficult for the message-building enzyme to get in and read the letters in the first place.

Precisely how stretches of non-coding DNA can help to regulate gene expression is the subject of a great deal of active research. We know that some of the non-coding regions produce RNA strands, and although these are never translated into proteins, they can interact with other molecules to repress or enhance the expression of certain genes and they are often found clustered around them. But much of the non-coding DNA seems to be true junk, simply clogging up the genome and serving no real purpose.

The inclusion of sections of nonsense *within* genes certainly complicates the manufacture of RNA messages. In bacteria, the ribosomes don't even wait for the RNA message to be completed before they attach themselves and begin their translation. In eukaryotes, this has to be prevented, as the initial RNA message is full of nonsense that needs to be carefully removed. Indeed, one of the main functions of the nucleus is to keep the ribosomes out: they are confined to the cytosol and have to wait for finished messages to be exported out of the nucleus before they can get to work on building the protein.

Inside the nucleus, an enormous cut-and-paste job is carried out on every RNA message, conducted by a huge piece of cellular machinery called the *spliceosome* (pronounced: *splice-oh-zome*). This acts like a ruthless film editor, cutting the junk out of the message, and stitching the important pieces back together again. It's hard to see what possible advantage such a ridiculous process could have, but it does give eukaryotic cells some unexpected creativity. The spliceosome can create slightly different proteins from any given gene by stitching the useful pieces together in slightly different ways. So, although humans 'only' have around 20,000 protein-coding genes, we can produce many more different proteins through imaginative cutting and pasting.

DNA sequencing has revealed other surprises about eukaryotic genomes, beyond the presence of non-coding DNA. Some of the genes in eukaryotic genomes are almost identical to bacterial genes – and it's not a random subset. The bacterial-looking genes are generally involved in metabolism, and many of the proteins coded for by these genes are sent directly to the mitochondria, the metabolic centre of the cell. We now know that mitochondria were once free-living bacteria and would have been engulfed with their own genomes complete and intact. But as part of the long process of enslavement, many of their genes have been moved to the host's genome, leaving them with just enough DNA to manufacture their own ribosomes and a few other essential components. The upshot is that mitochondria could never resume their free-living former existence – they simply don't carry enough information to

A Comparison of Bacterial and Eukaryotic Genomes

Typical bacterial genome

*blah*GENE*blah blah blah*GENE
*blah*GENE*blah*GENEGENE*blah*
*blah*GENE*blah*GENE*blah blah*

Typical eukaryotic genome

*blah blah blah blah blah*GENE
blah blah blah blah blah blah blah
*blah blah blah*GENE*blah blah*
blah blah blah blah blah blah

run their own affairs – a warning to us all that freedom of information is crucial for an independent existence.

So, the typical eukaryotic genome is certainly a strange affair. The hybrid ancestry of eukaryotes is very clear to see, with bacterial genes from captured metabolic slaves scattered throughout a genome that is otherwise more similar to archaea. Sprawling between and within these genes is a mass of repetitive junk, some derived from viruses, leading to enormous genomes that are coiled and packed so firmly that several metres of DNA can be squeezed into each tiny nucleus. In some multicellular beings, junk DNA seems to have got totally out of control, with less than 5% of the genome actually coding for proteins. It seems an odd start for creatures that we consider to be the ultimate life forms on our planet.

The giant amoeba, *Pelomyxa,* can be 5 mm long and contain multiple nuclei

When Darwin was developing his theory, he made it clear that evolution by natural selection was a slow process. He warned that species could only change gradually and by tiny increments, as if forced to take baby steps through an evolutionary landscape of possibilities. Certainly, he considered it quite impossible that a species could take a flying leap in the dark by making a sudden dramatic change that would actually work. Such 'Hopeful Monsters', as they were dubbed, were too unlikely to produce a viable creature, and anyone suggesting otherwise was subject to scorn and ridicule. This is partly why the idea of hybrid Hollywood creatures, like Sharktopus or Piranhaconda, are generally considered to be nothing more than the fantasies of feverish movie producers. But the monster that we might dub BacteriArchaea (aka eukaryote) is all too real.

Engulfing another cell and then appropriating not only its genome but all of its machinery in one fell swoop is truly an event so extraordinary that it's perhaps unsurprising that Lynn Margulis wasn't taken seriously when she first proposed that mitochondria were enslaved bacteria. But once accepted, other questions naturally arose. Was this THE defining event of the eukaryotic cell? After all, there are many differences between prokaryotes and eukaryotes, so perhaps the enslavement of mitochondria was a late event in their development. And when an extraordinary amoeba that lacked mitochondria was found in a pond outside the Oxford University Museum of Natural History, it seemed that this was a very real possibility.

At the museum, specimens arrive from all over the world in various stages of decay. First, they have to be prepared, and this often means removing the flesh of dead animals in order to preserve and display their bones. The story goes that the flesh of one Asian elephant was dumped in a pond near the museum in the 1970s and a student discovered a suitably gigantic amoeba living on its rotting remains. The most extraordinary feature of this amoeba was that it lacked mitochondria, and indeed the central warehouse for processing and delivering proteins. Surely then, this must be the most primitive eukaryote ever discovered?

Unfortunately, genome sequencing has revealed that the giant amoeba isn't primitive at all. Instead, the tell-tale genes from the ghost of an enslaved-mitochondria-past are clearly visible within its genome – despite the fact that actual mitochondria are now absent. We know that evolution doesn't always make organisms more complex and that apparently useful features can be lost (like the flightless birds on islands), but we still don't know why this giant amoeba lost its mitochondria, and with it the power of aerobic respiration. Intriguingly, it has other types of bacteria living inside its gigantic living space, which seem to provide it with essential metabolic services. But, it has clearly evolved down a rather special evolutionary path and is not the fabled early eukaryote that many biologists wished for.

Indeed, the sequence of events that led to the emergence of the modern eukaryotic cell is still shrouded in mystery. Perhaps it was a long and arduous journey and this is why it took eukaryotes the entire boring billion before they began to form multicellular beings. But eventually animals, plants and fungi emerged, and set the eukaryotes on a course for world domination. Well, that's our take on it, obviously. Bacteria no doubt have different stories.

So, now we come to the crunch. Eukaryotic cells didn't just stop once they were the biggest cells on the block – they went a step further and started to form new kinds of individuals built from millions of cells with a single-minded purpose. Cyanobacteria form multicellular strands because some cells can take on specialized chemical roles, like nitrogen fixation, that simply aren't compatible with other activities, like photosynthesis. But bacteria have a broader range of chemistry at their disposal (for example, no eukaryotic cell can fix nitrogen), so chemical specialisation is unlikely to have driven eukaryotic cells to working together.

Eukaryotic cells may have a narrower range of chemistry when compared to bacteria, but they have enormous flexibility in terms of shape and behaviour. This might allow them to form a larger structure with benefits for all, and we see this in certain kinds of amoebas called *slime moulds*. These amoebae spend most of their lives alone, happily engulfing bacteria in damp soil and leaf litter; but food isn't always plentiful, and if they begin to starve then something remarkable happens.

Just like the opportunistic bacteria that cause pneumonia, the starving amoebae start to release chemical cues into the environment. These cues say, 'I'm starving!' rather than 'I'm here!' and they cause other cells to move towards them and stick together to form free-flowing lines that converge on the single amoeba identified as the leader of the pack (usually the one that shouts, 'I'm starving!' the loudest). The enormous collection of around one hundred thousand cells then forms a 'slug' – usually around two to four millimetres long – and, incredibly, the 'slug' behaves as a single organism. It crawls off, seeking the right conditions, and when happy with its location, the slug starts to grow upwards to form a tiny mushroom-shaped structure, with a narrow stalk and a blobby head, called a 'Mexican hat'. Some of the amoebae migrate upwards and turn themselves into spores that can be launched from the dizzying height of one to two millimetres, allowing them to colonize new ground. And when conditions are right, the spores will germinate and the emerging amoebae can resume their lonely lives until the next episode of starvation brings them together again.

The Lifecycle of a Slime Mould

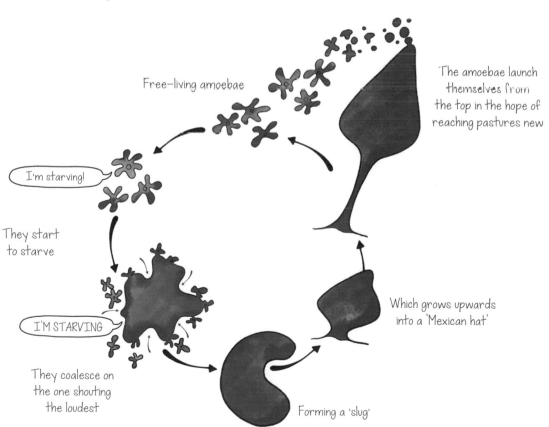

Free-living amoebae

The amoebae launch themselves from the top in the hope of reaching pastures new

I'm starving!

They start to starve

I'M STARVING

They coalesce on the one shouting the loudest

Forming a 'slug'

Which grows upwards into a 'Mexican hat'

By working together, the amoebae can colonize a much wider area than they could have managed alone. And, just like the cyanobacteria, cells within the larger structures have specialized roles: inside the 'Mexican hat' some cells form the stalk, and some the hat, and some move up and down between the two. But there is a major downside to coming together to make a larger structure. Some of the cells that form the stalk never get to launch themselves from the top of the hat and thus seem to have sacrificed themselves and their chance to reproduce to help others.

The fate of Galahad, that poor doomed wildebeest, has already revealed that co-operation between unrelated cells is difficult or impossible to achieve. The amoebae within the slime mould are probably related, but we know that collectively building a single-minded giant (like a human or a tree) requires an even higher degree of co-operation and it can only be achieved by a single cell and its descendants. The eukaryotes, with their enormous range of shapes and behaviours, are perfectly placed to sculpt such extraordinary beings, and although they took their time getting started, around 800 million years ago, they shifted up a gear and started to produce a dazzling variety of shapes and forms.

The emergence of true multicellular life transformed planet Earth. One of the most significant events took place around 541 million years ago, with an outpouring of animals so vibrant and so sudden that it still gives biologists headaches. Called the Cambrian explosion, it is one of the great events in the history of life, and it brought forth every modern group of animals on Earth today in the blink of an evolutionary eye.

Chapter 7
ANIMALS
The unstoppable rise

If we imagine squeezing the history of life on Earth into a single year, then we have reached 18 November, a momentous day, in which the eon of visible life is ushered in with a bang – possibly followed by whimpering – as this is the day that the animals first leaped onto Earth's stage. Although eukaryotic cells first appeared around 7 August, the months corresponding to the boring billion are now over, swept away by a sudden burst of multicellular creativity, the like of which has never been seen before or since on our planet. This concentrated flurry of evolution will give rise to every major group of animals, and their arrival will transform planet Earth, shaking up ecosystems and eventually spawning giants that will dominate both the oceans and the land.

During the boring billion, eukaryotic cells diversified into many forms and types, like the predatory amoeba or the ciliated *Paramecium*. Most of these prospered perfectly well as single cells, and they can still be found all over the world today, but a few gave rise to complex multicellular beings. The animals are one such success story and because they can all claim descent from the same single-celled ancestor, we consider them to be a unified group, referred to as the *Animal Kingdom*. So, who dwells within this realm and what features do they share?

First, all animals are multicellular creatures, made from many different cell types that work together seamlessly. Second, most of them can move – some with a grace and vigour that leaves other animals breathless – although there is a surprising number that spend their adult lives immobile. Third, nearly all animals reproduce sexually by producing egg and sperm cells. And finally, animals obtain the building blocks and energy they need by eating other organisms (often each other) and the energy contained within their food is released by aerobic respiration using the captured mitochondria within their cells.

If asked to rattle off a few names, a typical list of animals might include birds, mammals, lizards, frogs, fish, sharks and turtles. But all of these belong to the same major group within the Animal Kingdom and – according to most zoologists – there are around 35 such groups to choose from. The general name for one of these groups is a phylum (pronounced: *fye-lum*), and the plural is *phyla* (pronounced: *fye-la*). Each phylum is characterized by a unique *body plan*, defined by key features, like the number of body segments, the type of skeleton and the lines of symmetry. Humans, along with other mammals, birds, reptiles, amphibians and fish belong to a phylum called the *chordates* (pronounced: *cord-ates*), which contains around 70,000 species. But given that the best estimate of the total number of animal species on our planet is a staggering 7.7 million, then clearly, if we just focus on the chordates, we will miss most of the diversity on Earth.

Astonishingly, all animal phyla on Earth today that have a fossil record can be traced back to the same spectacular event: the Cambrian explosion. The Cambrian is the very first slice of time within the Phanerozoic, the eon of visible life, and it officially began on 18 November on our one-year timeline, or 541 million years ago. Within around 20 million years of its start-date, nearly all animal phyla alive today had sprung into being – and given that it took eukaryotes one billion years to even get started on building multicellular life, this is truly incredible. Even Darwin noticed that rocks older than the Cambrian appeared to be more or less devoid of life, while in younger rocks, a rich fossil record could be pieced together. Just how such a vast array of animal life could explode forth in such a short span of time puzzled Darwin throughout his life – and it remains one of the great mysteries of biology. So, what kind of eukaryotic cell might have been the ancestor of all animals?

Under the glare of a light microscope, a lone cell busily feeds on bacteria. It belongs to a group called the *flagellated funnel cells*, and unlike the shape-shifting amoeba, it doesn't need to chase down its bacterial quarry. Instead, like a miniature circus master, it reels them in by twirling an extremely long flagellum. The motion whips up a current of water that sweeps the bacteria inwards towards the funnel-shaped top of the cell, where they are trapped by a ring of long projections resembling a forest of tiny fingers – and once caught, the captive bacteria are engulfed, ready for digestion.

Flagellated funnel cells might seem like just another obscure eukaryote, but they are of particular interest to those who claim membership of the Animal Kingdom. Unravelling the origins of any group is hard, but we are fairly confident that flagellated funnel cells – or something very like them – were the forerunners of all animals. The unlikely journey that took these bacteria-guzzling cells from their lonely beginnings to the enormous diversity of multicellular animals that have creeped, crawled and cavorted their way round the planet for the last 540 million years or so is an incredible story of innovation and co-operation. But why do we think that these cells could be our ancestors?

Funnel cells normally use their long flagella to swim around freely, but in the presence of certain types of bacteria they have a very unusual trick up their sleeves. As they divide, the daughter cells don't fully separate, but instead remain connected by a slender bridge of cytoplasm. Fifteen hours later, and further rounds of cell division have produced a tiny *colony* of 10 to 12 cells, all joined together with their flagellae facing outwards. Tumbling through the water, the colony is driven by their combined waggling, and we believe that this strategy can deliver more prey to each cell than they would have obtained alone. This kind of close co-operation between a cell and its descendants is very similar to the co-operation we see within multicellular animals – with one crucial exception. In the absence of the right bacteria, the colonies split up and the cells go their separate ways, but the cells inside the bodies of animals don't have this kind of flexibility – instead, they are committed to the multicellular life.

The importance of flagellated funnel cells to the Animal Kingdom becomes clear when we consider one particular animal phylum – the *sponges* – widely believed to the be the first animal group to evolve. A sponge is filled with internal channels and chambers, and these are lined with *collar cells* that bear an uncanny resemblance to the flagellated funnel cells. The collar cells waggle their flagellae to create an impressive flow of water through the sponge body, and then trap and remove any bacteria that are swept within their reach, engulfing them individually and whole. But, unlike the free-wheeling funnel cells, there's no possibility of sponge collar cells returning to the single life – they can now only thrive as part of an impressive larger body.

Sponges can be found throughout the oceans of the world, wherever they can find a place to attach themselves. Usually this is a rock or the seabed, but sometimes they can be found growing on the hard parts of another animal, like the shell of a crab. Some live at great depths, where the water is cold and food is scarce, and these sponges grow extremely slowly and live to a great age, taking around one thousand years to reach their full height of up to three metres.

Funnel Cells vs the Sponge

Flagellated funnel cell

Sponge collar cell

Equipped with a long flagellum to reel in bacteria

Equipped with a long flagellum to reel in bacteria

Can be found swimming around freely

Or forming a small colony

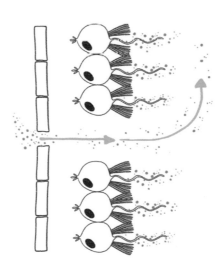

Can only be found inside the body of a sponge. The arrows show how water, carrying bacteria, flows through the sponge

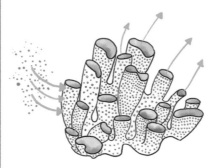

When certain types of bacteria are present, daughter cells stay together to form a multicellular body. But this is only a temporary arrangement

Collar cells line the inside walls of the sponge and create the water flow. They cannot leave and live independently

One of the reasons we believe that sponges were the first animal group to evolve is their simple construction. The walls of an adult sponge are built from two layers of cells: an inner sheet of collar cells, and an outer sheet of tightly packed brick-like supporting cells. Within the hollow walls, a third type of cell takes the form of amoebae that crawl around, engulfing stray bacteria and transporting food from the greedy collar cells to the supporting cells that don't get much chance to feed. But, these three cell types are pretty much all that the adult sponge consists of.

The adult sponge might be fixed to the spot, but it still has to reproduce. Sometimes small body fragments simply detach themselves and drift away, before settling down to form a new adult, but sponges – like other animals – also have sex lives. When ready to reproduce, some sponges cast both eggs and sperm into the water, but more often the eggs are held firmly within the body of the sponge, waiting for sperm from another sponge to arrive. When sperm cells from the right species enter, they are engulfed by the collar cells and delivered to the egg cells that wait within the sponge's walls. Here, the two cells fuse and develop into a *larva* – the pre-adult stage of many animals. The larva grows by cell division into a cylindrical blob that can swim away from the parent sponge in the hope of being caught by ocean currents and transported to a suitable place where it can settle down and build a new adult body.

To many of us, sponges may hardly seem to qualify as animals at all, but their lifecycles display many features that are absolutely typical of more familiar animals. Sponges feed by preying on other cells and using their digested remains to fuel their metabolism through aerobic respiration; the adults produce egg and sperm cells that fuse to form the first cell from which a new individual develops; and this individual builds an organized multicellular body using enormous numbers of cells of more than one type, all descended from that first cell. In fact, give or take a few important details, all animals that have ever lived are little more than glorified sponges.

Despite definitely being an animal, a sponge is a much looser collection of cells than the more integrated bodies of more familiar animals. Perhaps the biggest limitation of the sponge body plan is that the collar cells – just like the lone amoeba – are limited to engulfing single cells (mostly bacteria) and can't process larger items. But capturing and processing tiny individual food items is inefficient, so very few cells within the body can be spared for other roles.

The remaining animal phyla don't rely on ingesting single cells, but instead capture much larger items of prey. These are digested inside a dedicated cavity within the body, and the cells that line it only absorb the building blocks from which their prey are ultimately constructed, rather than engulfing whole cells, like the sponge. A key innovation for this dietary shift was the evolution of a dedicated leak-proof space inside the body, into which digestive enzymes could be poured onto the unfortunate victims. So, which animals first introduced this ground-breaking innovation?

On a sunny day at the seashore, it's easy to encounter a group of animals that appear to be made entirely from jelly. Stuck fast to the wall of a rock-pool, we might spy a sea anemone waving its menacing tentacles, or wander past a washed-up jellyfish lying forlornly on the sand. Jellyfish and sea anemones both belong to the same phylum: the fearsome *Cnidarians* (pronounced: *nid-air-ee-ans*) and they possess an extraordinary weapon – a venomous harpoon, which can potentially kill a human. Cnidarians array these weapons on their tentacles and different groups then deploy them in different ways. Some, like sea anemones and corals, stick one end of their bodies down onto a solid surface and point their weaponized tentacles upwards, while others, like jellyfish (definitely not real fish) float around with their deadly tentacles dangling downwards.

There is a second, lesser-known phylum of jelly-animals called the *comb-jellies*. Like jellyfish, these float around the oceans, but instead of firing venomous harpoons, they catapult tiny glue-covered nets to capture their prey and haul it in – but they are far too weak to capture a human. The comb-jellies' claim to fame is that they are the largest animal to propel themselves along entirely using cilia, which cover the outside of their bodies and are organized into structures called combs, which give these animals their name.

Whether they catch their prey with harpoons or nets, all jelly-animals are equipped with both a mouth and a leak-proof cavity to digest the prey once successfully captured. The great advantage to this strategy is that, unlike the sponge, jelly-animals don't have to force enormous volumes of water through their bodies in order to filter out enough food. Instead, they can grab any good bits that they encounter and pass them along to the mouth. Once inside, the cells that line the body cavity secrete enzymes to aid digestion. But forming a leak-proof cavity is no easy matter and requires the possession of specialized cells that can lock together so tightly that very little can pass between them.

In all animals except sponges, this digestive cavity, or gut, is lined by a sheet of specialized *epithelial cells* (pronounced: *ep-ee-thee-lee-al*) – a key animal innovation. Within the sheet, cells are packed together side by side and knitted to their neighbours – rather like a line of humans with tightly linked elbows – so that when one cell moves, the others move with it. To strengthen the sheet, each cell is firmly bound to a thin layer of material that underpins the cells and that they themselves secrete. Together, the tightly knit epithelial cells and their supporting membrane form a thin, flexible sheet that can move and stretch as water and food enter the body cavity – for example, when a greedy cnidarian opens its mouth.

Surprisingly, perhaps, jelly-animals are entirely constructed from just two layers of epithelial cells – one on the outside and one lining the body cavity – with jelly sandwiched in between. The jelly is mostly composed of water, but it also contains proteins with elastic properties, like *collagen*, that the epithelial cells secrete. In fact, secreted proteins are a crucial feature in the construction of many animal bodies, with hair, scales and feathers being more familiar examples.

Epithelial cells aren't the only innovation to be found in the jelly-animals. They also contain two other specialized cell types that are found in all animals except sponges: *muscle* and *nerve cells*, both absolutely essential for any animal that wants to move and do so in a co-ordinated way, rather than just hopelessly wobbling like a – well – like a jelly.

The powerful and sometimes balletic movements of many animals depend entirely on muscle action. Unlike sponges, the jelly-animals possess simple muscles, and although they are formed in a rather different way from the ones in our bodies,

The Construction of a Sea Anemone

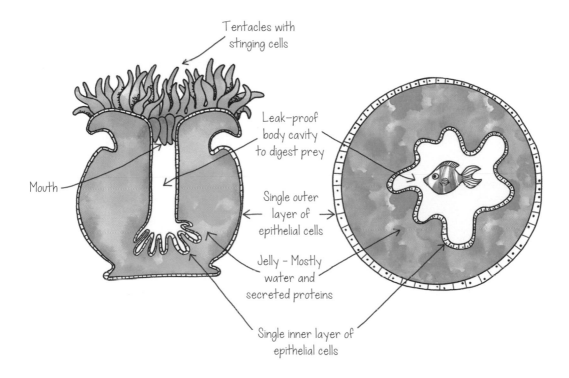

Tentacles with stinging cells

Leak-proof body cavity to digest prey

Mouth

Single outer layer of epithelial cells

Jelly – Mostly water and secreted proteins

Single inner layer of epithelial cells

their mode of action is identical.

Although there are different types of muscle, the basic principle of muscle contraction is always the same. Muscle cells contain filaments of the cytoskeleton laid out in regular parallel rows, with thick filaments interspersed among thinner ones. The thick filaments can slide between the thin ones, and you can see how this works by holding your hands out in front of you, palms facing towards you with fingertips touching and thumbs pointing upwards. Your thumbs represent the two ends of the muscle cell and your fingers represent the filaments within. Now, open your fingers, and allow the fingers of one hand to slide between the fingers of the other. You should see that your thumbs have moved closer together, and this is how the sliding of filaments within muscle cells causes the cells to shorten and the muscle to contract.

In a muscle cell, the thick filaments have blobs protruding from the sides that can ratchet along the thin filaments. This movement uses up an awful lot of ATP batteries, but once contracted, the thick filaments can't move themselves back again — it's as if the fingers of your two hands were now trapped together. The unfortunate consequence of this arrangement is that the muscle will remain in its contracted state unless something pulls the thick filaments back to their starting positions, a job normally performed by a second set of muscles.

To see how this works, let's consider a simple cnidarian called *Hydra* that looks rather like a sea anemone, but can be found in ponds and streams. Imagine a thin hollow cylinder closed at one end and stuck down to a rock. The open end of the cylinder is the mouth, surrounded by a ring of armed tentacles, which closes tightly when newly caught prey has been transferred inside for digestion. The walls are made from two thin layers of epithelial cells with jelly in between, and some of the epithelial cells have woven their tails together to produce two sets of muscles within the walls.

The first set of muscles stretch from the top to the bottom of the cylinder (or base to tentacle), while the second, so-called *circular muscles*, form rings around the body. If *Hydra* wants to make itself short and fat, then it contracts the first set of muscles, which will shorten the distance from bottom to top — but without the second set of muscles it would be stuck in this squat state forever. Fortunately, when the circular muscles contract, they squeeze the body inwards, making it long and thin again and so stretching out the first set of muscles. So, by judiciously contracting the two sets of muscles, *Hydra* can be as long and thin (or as short and fat) as it likes.

The possession of two sets of muscles with opposing effects is a common feature of most animals — but there are some exceptions. One is *Hydra's* floating cousin, the jellyfish, which has powerful muscles that circle the bell at the top of the animal, from which the tentacles hang down. Contraction of the circular muscles constricts the bell and forces water out, causing the jellyfish to whoosh gently forwards under a slightly feeble form of jet propulsion. But when the circular muscles stop contracting, the bell naturally springs back to its starting position without the need for a second set of muscles. This energy-saving device is due to the remarkable elastic properties of the jelly with which jellyfish are filled — and it's yet another hats-off success for the canny Cnidarians.

The Muscles in Hydra

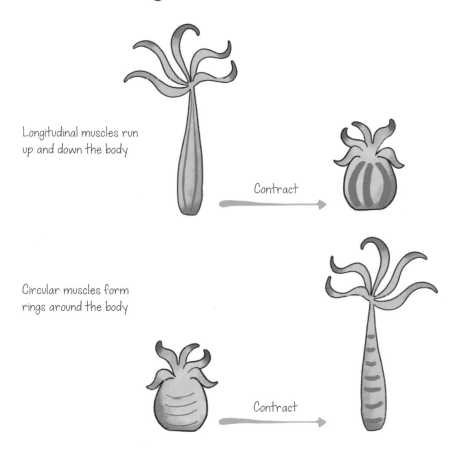

Longitudinal muscles run up and down the body

Contract

Circular muscles form rings around the body

Contract

Muscle contractions need to be co-ordinated or some hapless *Hydra* might try to make itself long and thin AND short and fat all at the same time, which would use up ATP batteries and achieve absolutely nothing. To prevent this kind of mishap, muscle action is co-ordinated by *nerve cells*, yet another type of specialized cell that first appeared in jelly-animals. Nerve cells can take a great variety of shapes, but they usually have a round cell body, where the nucleus resides, and one or more long thin projections that reach out and contact other cells, such as muscle cells.

To make the muscle contract, nerve cells send an electrical signal that travels very quickly along the cell membrane. Once the signal reaches the end of the cell, it triggers the release of chemicals called *neurotransmitters* that diffuse across the tiny gap between the nerve cell and the muscle cell. When the chemicals arrive at the muscle cell they bind to receptors in the membrane, triggering a series of events that cause muscle contraction. But *Hydra* doesn't want to be at the mercy of overly twitchy nerve cells and waste precious ATP batteries contracting its muscles without good reason. So how do the nerve cells know when to send a message and when to stay quiet?

In a jelly-animal, some of the nerve cells act as sensory cells, and they detect different stimuli, such as light, touch or the presence of certain chemicals. The detection of the right stimulus causes the nerve to fire off an electrical signal, and because each nerve cell is connected to others, the whole animal can make a co-ordinated response. Admittedly, *Hydra* has a fairly limited range of behaviours: it can retract itself into a tiny blob; stretch itself up to its full height; extend and wave its tentacles; or pull them towards its mouth once it has caught something – hardly the stuff of nature documentaries, but it does have one rather fantastic trick up its sleeve.

Hydra Performing a Cartwheel

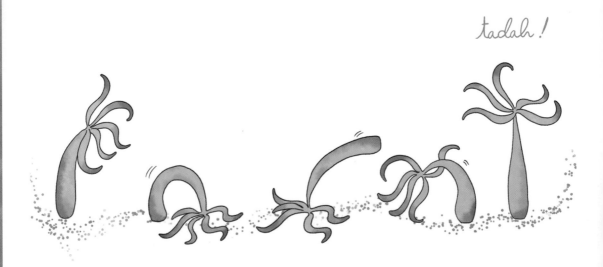

If *Hydra* are unhappy with their location, they can move away rapidly by performing a cartwheel. First, they bend themselves over sideways until their tentacles make contact with the ground. This allows them to unstick their bottoms and flip themselves base-over-tentacle until their bottom end is back on the ground again – and once safely stuck down they release their tentacles and stand up straight. Resisting a natural desire to applaud this tiny gymnast, we can certainly say that muscle and nerve cells really do make a difference to the life of an animal.

For many of us, the jelly-animals probably feel closer to 'real' animals than the poor overlooked sponge; but they still don't cut it for many humans. So, what do most of us consider a 'real' animal to be? Animals with fur or feathers and big eyes tend to top the lists of favourites, so a tiger or an eagle would certainly qualify, maybe even a ladybird or a snail. But what about a lowly worm? In fact, all animals that are neither sponges nor jelly-animals belong to the last great group of animals: those with bilateral symmetry.

Just about every familiar animal on our planet belongs to a huge group containing multiple phyla called the *Bilateria* (pronounced: *By-lat-ear-ee-a*). The name was chosen because they have a plane of symmetry that cuts down the middle of the body and gives them (more or less) identical left and right sides. But, there are no other planes of symmetry, so the back is different from the front and, crucially, the head is different from the tail. Indeed, a better way to think of bilaterians is simply as animals with clear front and back ends.

The distinctive front and backness of bilaterians is driven by yet further improvements to their digestive systems. It's embarrassing for Cnidarians, but despite all their many innovations, they do not possess an anus. Instead, they have to open their mouths to rid themselves of undigested bits of food and then flush out their body cavities with clean water. But nearly all bilaterians are spared such indignities because they possess what is known as a *through-gut*: a narrow channel lined with epithelial cells that runs through the body from the mouth at one end to the anus at the other, although comb-jellies – to their great relief – have recently been shown

The
box jellyfish
has a
cluster of
six eyes
at each
of its four
corners

to have the same sensible construction.

The second major difference between bilaterians and the jelly-animals is that the space between the outer and inner epithelial sheets is no longer filled with jelly. Instead, bilaterians possess a third layer of cells that develops into solid blocks of muscle that power the body forwards, with the head leading the charge and the tail reluctantly bringing up the rear. To aid this purposeful movement, the bodies of many bilaterians are divided into segments, from which useful things – like bristles, legs or even wings – sometimes sprout.

The dogged determination of bilaterians to meet life head-on has spurred the evolution of sense organs, like eyes and antennae, that are concentrated around the front end. Highly developed eyes that can form an image – rather than simply detect light – can be found on many bilaterians as far back as the early Cambrian, allowing animals to see one another for the first time, and then presumably use their newly evolved locomotive powers to either give chase or take evasive action. But, while the movements of *Hydra* are somewhat coarse and clumsy, bilaterians must act on faster timescales and exercise finer control if they are to successfully catch their dinner or avoid becoming someone else's.

To achieve better co-ordination, bilaterians have centralized their nervous systems by bundling nerve cells together into a *nerve cord* that runs from front to back. In most bilaterians the nerve cord swells at the front end into some sort of *brain* where information is integrated and decisions made – potentially leading to more complex behaviours. As part of this reorganization, nerve cells – or *neurons* – in bilaterians come in two main flavours. *Sensory neurons* carry information from sense organs, like eyes, to the central nerve cord – where they can be rapidly transmitted to the brain. Once the brain has made its decision, the call to action is quickly relayed back along the nerve cord to *motor neurons*, which fire off messages to the appropriate muscles, causing them to contract and effect a co-ordinated response.

Firm fossil evidence of the earliest bilaterian currently dates to around 12 million years *before* the official start of the Cambrian, although – given the endless appetite of fossil hunters for new finds – don't be surprised if even earlier bilaterians are soon found. The fossil in question is an amazingly rare find – a perfectly preserved soft-bodied animal that has left a clear trail behind it, revealing its final movements. It's unquestionably a bilaterian – an incredible 27 centimetres long – with a segmented body that had allowed its owner to wiggle its way across the sediment before dying. But it's spectacularly good evidence that the Cambrian explosion might have started a little further back in time than we had originally thought.

So, who exactly are the bilaterians? Zoologists argue over the number of bilaterian phyla that they are prepared to recognize, although most agree it's somewhere between twenty-five and thirty. Most of these are different types of worm, many of them tiny and only found in the oceans, but we shouldn't underestimate their importance. Worm isn't a technical word and can be applied to any bilaterian animal, longer than it is wide, which doesn't possess proper legs. But the grandest worm phylum is undoubtedly the *Annelida* (pronounced: *Anna-leed-a*), a group containing some familiar friends – like the humble earthworm – and some beautiful, if bizarre, strangers.

Watching an earthworm move across the surface of the soil is a fascinating sight, available to view at absolutely no cost in any good garden or park. The body of the earthworm is divided into segments, each equipped with muscles that work in opposition, like those in the body of *Hydra*. Each segment has both circular muscles that form rings around the body and a second set that run parallel to the body axis, called *longitudinal* muscles. Squeezing the circular muscles makes the segment long and thin, while contracting the longitudinal muscles makes the segment short and fat. But, if the worm contracted the same set of muscles in all segments at the same time, it would just alternately get longer and then shorter again, but wouldn't achieve any forward motion. To get the worm moving, the segments co-ordinate their contractions and make use of short bristles that protrude from the segments. Some segments become short and fat, pushing the bristles into the ground for anchorage, while others concentrate on extending forwards – giving the worm a fighting chance of escaping a pursuing bird, no matter how early it turns up.

Annelid worms are spectacularly successful, with around 22,000 species currently

Earthworm Locomotion

An earthworm without segments can't achieve any forward motion

It can use the longitudinal muscles to become short and fat ...

Contract

And it can use the circular muscles to become long and thin

Contract

Segments and bristles allow it to move forward

Each segment contains both types of muscle. When the longitudinal muscles contract, short bristles anchor the worm's body against the ground

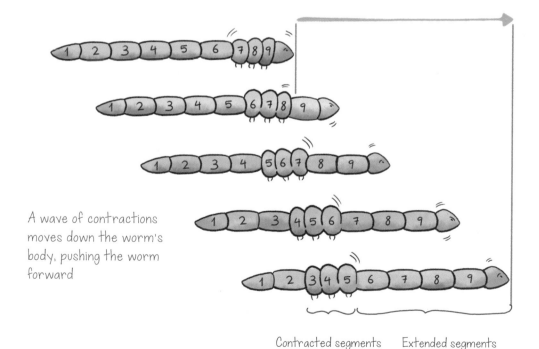

A wave of contractions moves down the worm's body, pushing the worm forward

Contracted segments   Extended segments

described, most of them marine. Many burrow into soft sediments, where they feed on detritus; a few are equipped with jaws to grab unsuspecting passers-by; but most *filter feed*, a technique that involves filtering out any food from the water using specially adapted mouthparts. Some of the filter feeders – like feather-duster worms – possess elaborate crowns of feathery mouthparts covered in cilia, which they warily protrude from their burrows in order to trap anything small enough to eat. These delicate structures are a potentially delicious snack for passing predators, so feather-duster worms have evolved an astonishing array of eyes, including a pair of gigantic compound eyes with multiple lenses that are specially attuned for spotting approaching fish. The arrival of such a predator causes the feathers to be retracted at lightning speed, a trick enabled by the annelid equivalent of superfast broadband: giant nerve cells that run the entire length of the body and transmit signals exceptionally quickly.

Feather-duster worms might have lightning reactions, but worms are best adapted for burrowing through sediment or soil and much less efficient at getting from A to B. For moving across a surface (like the seabed or a lawn), worm locomotion is fatally flawed, because the worm's whole body drags along the ground. Much greater efficiency can be achieved by sprouting legs, which greatly reduce friction and allow some animals to scurry away so rapidly that they induce panic in human observers. And one particular phylum has been accused of having far more legs than many humans would consider either necessary or reasonable.

Success can be judged in many ways. We might call an animal phylum successful because it has survived for a long time – in which case, the sponges should take the crown, as they probably evolved first and are still with us. But, many biologists consider a better measure of success to be the number of species that a particular group has produced – and in that case, there is only one winner.

Stepping up to receive the prize are the *arthropods*. This phylum includes insects, crustaceans (shrimps, crabs and lobsters), spiders, centipedes, and many others that we might call creepy-crawlies. The success of the arthropods is hard to overstate.

They make up around 80% of all described species on Earth and numerically dominate both the oceans and the land – with the insects being one of only four groups to have mastered powered flight. We probably all encounter several different arthropods every day, whether we notice them or not, and they play nearly every imaginable role within ecosystems, from prey to predator to parasite – although they don't photosynthesize, so like all animals, they are ultimately dependent on plants and other green things.

Key features that arthropods share include a body divided into segments protected by an *exoskeleton* – which is why most arthropods crunch when you (accidentally) step on them. An exoskeleton is rather like a suit of armour, offering protection for the soft stuff inside, but also a hard surface to which important things like wings and legs can be attached. All arthropods have jointed legs, allowing them to make sophisticated and precise movements, but the number of legs varies among the different groups, from the standard three pairs possessed by all insects, to the 200 pairs gracefully co-ordinated by some millipedes (and if you're wondering what the difference is between a centipede and a millipede – millipedes have two pairs of legs attached to each body segment, while centipedes only have one).

Variation in the number of legs is just one example of the extraordinary flexibility of the arthropod body plan, and this is probably key to their success. We can speculate that the ancestral arthropod might have had a large number of body segments, each possessing one or two identical pairs of simple legs. But, over time, these legs could become specialized for different functions, rather like an elaborate Swiss army knife. The crustaceans exemplify this very well – crabs and lobsters have ten pairs of legs, but the front pair have been adapted into menacing claws that can fight off enemies and rivals, while in some crabs, the last pair forms flattened paddles to aid swimming (which still leaves them with three pairs to walk around on).

Crustaceans are the most successful arthropods in the sea, although the tiny ones are far more important to ocean food chains than the more visible crabs and lobsters. *Krill* are small shrimp-like animals that occur throughout the world but are particularly abundant in the Southern Ocean, which encircles Antarctica. In these cold and stormy waters there is a single dominant species of krill with an estimated mass

The mass of krill in Antarctica is similar to the mass of humans on Earth

of around 380 million tonnes (comparable to the estimated mass of humans in 2019). Krill filter feed using modified legs to sieve tiny organisms from the water, and these are in turn scooped up by giant filter feeders, like the baleen whales, including the blue whale – the largest animal to have ever lived on Earth.

While crustaceans dominate the oceans, the insects take their place on land. Like their marine cousins, adaptable body parts are key to insect success. But while some have modified their legs for digging or jumping, the diversity of their mouthparts is truly extraordinary, allowing them to eat almost anything. Beetles have large chomping jaws that enable them to chew through all manner of tough stuff, including wood; butterflies and moths have a long, coiled proboscis to sip nectar from the deepest flowers; while many bugs have piercing mouthparts which they use to suck the sap of plants – or the blood of vertebrates. The widely disliked *mosquito* sucks small amounts of human blood – so little, indeed, that it should be nothing more than an annoyance – but, unfortunately, it also plays host to the parasite that causes the deadly human disease *malaria*, and by transporting the parasite around, it is currently held responsible for around 400,000 human deaths every year.

Finally, a few insects have formed complex societies, where only some individuals reproduce. Perhaps the most familiar social insect is the honeybee, where a single queen lays all the eggs. Most of these will hatch out as female workers that will never lay eggs themselves; instead, they will spend their lives gathering nectar and pollen to feed and rear their sisters. Some of the queen's eggs will develop into new queens and males, that will mate and form the next generation, and they too will be reared by the workers. The worker bees help their mother, instead of rearing their own offspring, for the same reasons that cells co-operate to form multicellular bodies: the bees within a hive are closely related, and by co-operating and dividing up tasks, they can achieve much higher fitness than by going it alone.

Some of the most sophisticated insect societies are found among a group called *termites*. Termites are closely related to cockroaches and are mainly found in tropical regions. Worker termites can be male or female, but like their honeybee counterparts, they do not reproduce. In addition to workers that gather food, some of the queen termite's eggs develop into fearsome soldiers, who, like the workers, are sterile. In some termite species, the soldiers have evolved a spectacular defensive weapon: a type of glue gun that they can aim at insect attackers, sticking them to the ground. The evolution of this gun has cost the soldiers their jaws, so they need to be fed by the workers or they would starve.

The advanced division of labour seen in the social insects, where individuals take up specialized roles, is a product of their tight co-operation. Indeed, a beehive or a termite colony can be considered as a kind of super-organism, where the individual is sacrificed for the greater good of the genes that they share (just like the cells in your body).

So, the arthropods are an incredible evolutionary success story. They have colonized every habitat on Earth, creeping and crawling their way into the deepest ocean and onto the highest mountain, eating everything in their path. Surely, no phylum can outdo them. But there is one thing that the arthropods haven't produced – an intellectual giant. And we don't need to visit our own phylum to find one.

Arthropods are just one of many groups that are sometimes called *invertebrates* to distinguish them from vertebrates (animals with backbones). But this is an extraordinary piece of conceit. The vertebrates are just one part of one phylum – the chordates, so they really don't deserve such special treatment. Indeed, arthropods and jelly-animals are just as different from each other as vertebrates are from arthropods, so to call something an invertebrate reveals virtually nothing about it – other than the rather uninspiring fact that it doesn't have a backbone.

The last great phylum of invertebrates is the *molluscs*. These are exceptionally diverse, and it's hard to believe that they belong together in one group and share a common ancestor. There are three main types of mollusc: the snails (with a single coiled shell), the *bivalves* (with two shells, like an oyster or a scallop) and the squid/octopuses (collectively known as *cephalopods*, meaning 'head foot', as their arms seem to sprout directly from their heads). And this last type of mollusc is the closest thing we have on Earth to an alien intelligence.

Stories of intelligent octopus abound. When kept in aquariums, octopuses soon get bored and need stimulation – new toys and new additions to their environment, which they enjoy rearranging. A few octopuses have found celebrity and caused headaches for their keepers: one sneaked out at night to pilfer crayfish from a neighbouring tank – naturally not forgetting to close both the lid of the crayfish tank, and its own, to cover its tracks. Otto, a resident of an aquarium in Germany, clearly didn't like having a light visible at night and squirted water at the lights in his tank every night, fusing all the lights in the building and baffling the aquarium's employees, while Sid from New Zealand tried so hard to escape and return to the ocean that the aquarium relented and took him back there themselves. So, why are octopuses able to succeed, when let's face it, a fish like Nemo would almost certainly fail?

Of course, when it comes to escapology, octopuses have a couple of major advantages over Nemo. They can survive a considerable amount of time out of the water and can manipulate their environment with superb aplomb: having eight arms

An Alien Intelligence

to juggle things with (and yes, Otto has been observed juggling hermit crabs when bored). But, their feats are also undoubtedly due to their intelligence. In recent years, octopuses have begun to fascinate humans – spawning books and documentaries – and many of these focus on human–octopus interactions. It's unclear whether they really are any more intelligent that many other non-human vertebrates, but perhaps there is something special about connecting with an animal that feels so different and alien compared to our human selves and with whom we last shared a common ancestor way back at the start of the Cambrian.

Despite our admiration, octopuses have been on the menu for thousands of years and are still a prized delicacy in many countries. More recently, fisheries have begun to target squid, which (like jellyfish) now occur in enormous numbers, as we have decimated their vertebrate competitors. Along the coast of South America, up to one million tonnes of Humboldt squid – a cephalopod monster that can reach 40 kg and pose a threat to human divers – can be caught each year, especially during years when warmer waters allow the squid to live longer and reach colossal sizes.

We have a fascination with intelligent animals, but we shouldn't forget that other molluscs also matter to humans. The oyster – now associated with fancy restaurants – used to be the food of the poor in many European countries, before human activity destroyed oyster beds, and their filter-feeding activities helped to keep coastal waters sparkling clear. Indeed, the reason that eating bivalves, like oysters or scallops, can lead to illness is entirely due to their tendency to consume anything and everything that the water contains – including bacteria that are harmful to humans and toxin-filled algae. Snails represent a safer bet, as they do not filter feed, but graze using a highly modified tongue, called a *radula*.

The molluscs are our final stop in this brief tour of the invertebrate animal phyla. They exemplify the extraordinary diversity that the Animal Kingdom holds, and amazingly this diversity did not take hundreds of millions of years to evolve. The fundamental body plans represented by the different phyla were generated in a burst of evolution that took place more than half a billion years ago and lasted for only 10 or 20 million years. So, what do we know of the Cambrian explosion and, crucially, *how* do we know it?

Our knowledge of this critical geological period comes from a few sites with exceptional preservation, in which the hard parts of animals can be clearly seen, but also – very often – their soft tissues. The first Cambrian site to be discovered was the Burgess Shale in Canada, but others have followed, including two incredible new sites recently uncovered in China: Chengjiang and Qingjiang. Comparisons reveal slightly different worlds and remind us that the number and type of animals found in oceans around the world varied just as much during the Cambrian as it does today.

All Cambrian sites host a bewildering variety of animal life, including hundreds of bilaterians fully equipped with heads, legs, eyes, antennae and extraordinary mouthparts. There are many unusual arthropods – relatives of today's insects and crustaceans – but most are different enough that we can't confidently assign them to a modern group. Annelid worms also abound, together with their tell-tale burrows and tracks, and there are even small chordates, around two to three centimetres long, that resemble small fish (if you squint at them in the right way). But the bilaterians don't have it all their own way: the Qingjiang site contains perfectly preserved specimens of a whole variety of jelly-animals looking just as fearsome as they do today.

Some of the Cambrian animals are huge, including the almost fantastical *Anomalocaris*. This beast had the inconvenient habit of falling apart when it died, and its scattered pieces were originally ascribed to a sponge, a jellyfish and a crustacean-like creature. Eventually, a whole animal was pieced together and the reality of what might be described as the first large predator finally emerged: an actively swimming proto-arthropod with huge eyes and a circular mouth, bearing a pair of long downward-curving appendages on its head – quite unlike any modern animal. It might have preyed on the abundant *trilobites*, another group of early arthropods that became hugely successful in the periods that followed the Cambrian, or perhaps it simply used its circular mouth to hoover up soft-bodied creatures hiding in the sediment. But, whichever food it preferred, *Anomalocaris* was the first in a long line of fearsome active predators – including *T. rex* and great white sharks – that have unconsciously shaped ecosystems, even as they have terrorized their victims.

Perhaps unsurprisingly, given some of the actors, many of the characters in the

The Great Predator of the Cambrian Seas

Is it a jellyfish?

Or perhaps the leg of a bottom-feeding scavenger?

No it's *Anomalocaris*! The world's first top predator

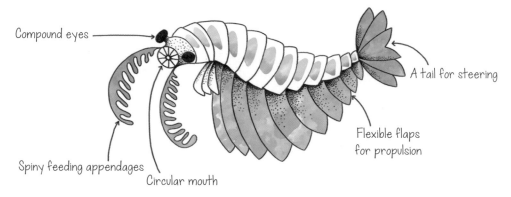

Compound eyes

A tail for steering

Flexible flaps for propulsion

Spiny feeding appendages

Circular mouth

Cambrian theatre are heavily armoured. Indeed, body armour swept through animal groups like some hot new fashion: arthropods toughened themselves by hardening their exoskeletons – as seen in the shells of modern crabs and lobsters, and deployed to great effect by the trilobites. Molluscs used the same trick to produce a hard shell into which they could retract their soft bodies, and there are many other fragments of shells and tubes in Cambrian deposits whose owners simply cannot be identified. So, is the Cambrian explosion real? Or is it just an explosion of body armour?

Today, most biologists believe that the Cambrian explosion is real and does require a special explanation. One possibility is that the evolution of the first mobile animals reshaped the world and provoked further rapid bursts of evolution. To examine this idea, biologists have turned to a few rare sites that offer tantalizing glimpses of life just before the Cambrian began. One of these is the Ediacaran

Hills in Australia, and it portrays a peaceful and unhurried world – very different to the bustling chaos of the Cambrian.

The world of Ediacara is filled with very strange soft-bodied creatures that are multicellular but often look nothing like anything alive today. Most of them don't appear to be mobile, but are upright frond-like structures with simple geometry that were fixed to the bottom of the sea. We don't know how they obtained energy – perhaps they photosynthesized, or trapped and engulfed bacteria in a similar way to a sponge – but their lack of modern counterparts means that we are always left guessing. Between the fronds, the sediments were covered in microbial mats, consisting of bacteria, archaea and single-celled eukaryotes, which could grow unhindered because there were no grazing animals to eat them. But this all changed as the Cambrian dawn drew near.

In sediments just before the base of the Cambrian, we see a world in crisis, as the Ediacaran idyll came under sustained attack. Fossilized tracks and burrows reveal that newly evolved bilaterians were munching their way through the signature microbial mats, triggering irreversible changes to their foundations. Ediacara had previously been undisturbed and very little oxygen would have penetrated into the sediments, but the arrival of determined burrowing animals changed all that, and opened up the sea floor to more and more inhabitants. Careful examination of the burrows and tracks has revealed that animals tunnelled deeper and turned more sharply as the Cambrian advanced – and these changes go hand in hand with an increase in body armour, as predators began to emerge that could feast on anything soft-bodied and immobile. In a relatively short space of time, the peaceful static world of Ediacara was gone, replaced by a dynamic ecosystem that teemed with animals very closely related to those that we know today.

Perhaps unsurprisingly, the causes of the Cambrian explosion are still hotly debated. Some biologists believe that an external trigger, such as a rise in oxygen levels, lit the fuse. Oxygen levels increased prior to the Cambrian – jumping from 0.1% to 1% of current levels around 800 million years ago – but it's harder to directly link further rises in oxygen to key Cambrian events. Others consider that the explosion was driven by evolutionary innovations: the ability to move, sense other animals, and digest larger prey may all have driven further dramatic expansions in animal form and function as they battled it out in an escalating arms race. But, in truth, it's always difficult to know precisely which factors contribute to any event.

Imagine that we were asked to explain why human technology suddenly exploded during the 19th and 20th centuries. Many different plausible explanations could be advanced – the invention of steam power during the Industrial Revolution surely paved the way for other, more advanced technologies – but such an invention required previous knowledge, garnered over centuries. Perhaps others could argue that better communication was more important than any specific invention, so that new ideas could be spread around the world, much faster than ever before. But if it's hard to agree on the underlying causes of a recent period in human history, it's inevitably more difficult to get to grips with a biological event that lies 540 million years in the past and where the evidence is limited to a few fossilized snap shots of its inhabitants.

Scientists may still argue over the causes of this ancient flourishing of animal life, but they all agree that the Cambrian is a pivotal moment in Earth's history. If we could travel back in time and visit the Earth at any point before the Cambrian, we would be greeted by a world so alien that we would probably think we had been transported to a different planet. But after the Cambrian, the world would feel familiar, at least in the sea. The actors might look slightly strange, but we would surely recognize that this was Earth: a teeming world of animals that crept, crawled, swam and hunted each other, aided by sense organs, powered by muscles, and co-ordinated by nerves. Since the Cambrian, animals have continued to evolve and change, but the fundamental character of the world they created was fixed around 540 million years ago and is still available to view in today's oceans.

Although the chordates appear to be minor actors in the Cambrian play, we know that this phylum includes the vertebrates, and that these animals went on to be highly successful. We are vertebrates ourselves and so naturally have a vested interest in knowing more about how vertebrate bodies are built (and surely octopuses would be just as self-interested if they ever took to writing books). There might have been tiny chordates in the Cambrian, but the defining feature of vertebrates once they rose to prominence was their size. So, how do we build a true giant?

Chapter 8

VERTEBRATES

Larging it – how animals grew up

Every animal, whether ant or octopus, belongs to one of the phyla within the Animal Kingdom, and we are no exception. Our home phylum is the chordates, which contains vertebrates (animals with backbones) and, strangely enough, some weird bag-like animals called sea squirts that stick themselves down on to rocks in the oceans and filter feed (and yes, if squeezed, they will squirt water at you). It's reasonable to ask what we could possibly have in common with such unlikely creatures, but as in many human families, it seems that close relations can continually surprise us with how different they are to ourselves.

Animals are placed in the same phylum if they share an exclusive common ancestor that *isn't* the ancestor of any other animal group. For example, we believe that sponges are the oldest animal phylum. If true, this means that the common ancestor of the sponges split away from the ancestor of all the remaining animal phyla some time before the Cambrian explosion, and so sponges can claim this ancestor as exclusively theirs.

Probable Relationships Within the Animal Kingdom

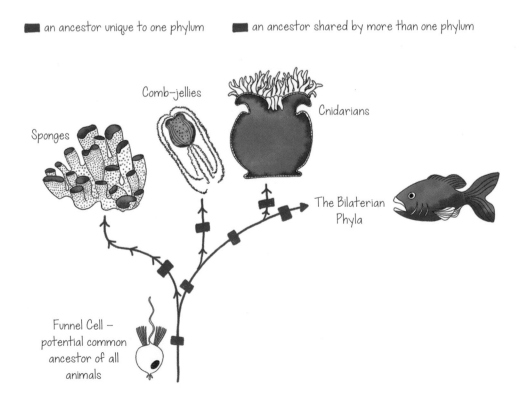

■ an ancestor unique to one phylum   ■ an ancestor shared by more than one phylum

Comb-jellies

Cnidarians

Sponges

The Bilaterian Phyla

Funnel Cell –
potential common
ancestor of all
animals

Sharing an exclusive common ancestor should ensure that modern members of the same phylum look alike, but although it's pretty easy to identify a sponge, it can be challenging to correctly identify which group some other animals belong to. The problem occurs because the common ancestor is long gone, and evolution has had a very long time to tinker with the features of its descendants. We already know that features can be lost as well as gained, such as powered flight. Nevertheless, it's easy to see that penguins are birds (and belong in the chordates) because they still have wings, even though they no longer use them to fly through the air. But if a feature was lost millions of years ago, then no traces may be left, or natural selection might have modified the original feature into something unrecognizable, making the task of assigning animals to the correct phyla difficult.

Since the Cambrian, natural selection has tinkered so heavily with chordates that their modern representatives seem to share very little in common. For this reason, biologists usually focus on what the chordate ancestor might have looked like and then note how its features have been modified in different types of chordate, like humans or sea squirts. So what features did the common ancestor of chordates possess?

In humans, two defining chordate features are closely intertwined within a single structure. The spine serves two important functions – it provides support and also protects a bundle of nerve fibres that carry messages between the brain and other parts of the body. The chordate ancestor, together with some modern chordates, didn't have a spine; instead it had two separate structures: a stiffening rod along the back to provide support and a bundle of nerves that ran alongside it.

The third common feature of chordates is the blocks of muscle that develop in the body wall on either side of the stiffening rod. These muscles are what we eat when tucking into a fish and are used to power the fish's body through the water. Muscles are a key feature of all vertebrates, although mammals use the muscles in the body wall to assist breathing, while relying on the highly developed muscles in their limbs to run, jump and swim.

Finally, the chordate ancestor had multiple slits in the side of its neck. These slits were used to filter feed, and sea squirts continue to use them for this purpose today. Water is sucked in and passed through the slits, where particles of food are filtered out before the water is expelled. In vertebrates, natural selection has repurposed these slits and their supporting arches. The first arch forms the jaws, while in fish the remaining arches support the *gills*, which are used to gain oxygen from the water. Land-living vertebrates have lost the gills and retained the jaws, but we can still see the bulges of lost gill arches in developing human embryos, allowing us to make the link to our past lives.

Sea squirts would never have been assigned to the chordates had their larvae not been found. Adult sea squirts don't have a stiffening rod or blocks of muscles in the body wall – indeed they don't even seem to have a head and a tail. But their larvae have all of these things and swim around, like small tadpoles, until they find a place that suits them, at which point they stick themselves down and transform into a filter-feeding bag.

The most well-known group of chordates is the vertebrates, which include various types of fish as well as turtles, lizards, frogs, crocodiles, birds and mammals. Vertebrates are defined by their backbone, although in sharks and rays this is not made of bone, but instead their entire skeletons are composed of a flexible, lightweight alternative called *cartilage*. While sharks have undoubtedly been successful, the biggest group of vertebrates on Earth today are the bony fish. Bony fish can be split into two groups: those with a series of fine bones in their fins that spread out like the rays of the sun (the ray-finned fish) and those with more substantial arm bones (the lobe-finned fish). Perhaps surprisingly, the vast majority of lobe-finned fish on today's Earth are not to be found in the oceans but on land. Called *tetrapods*, they are the descendants of a pioneering lobe-finned fish that found its way out of the water and founded a highly successful dynasty of vertebrates.

Tetrapods are vertebrates with four limbs. To find out how they evolved, we need to start where they began: in the oceans with their fishy ancestors. The very first fish swam in the Cambrian seas, but they were small and probably rather unimportant compared with arthropods like trilobites. But as the Cambrian ended, fish came into their own and began to increase in size and diversity.

The Chordate Ancestor and Two of its Descendants

Ancestral Chordate

Stiffening rod

Nerve cord

Tail

Mouth

Blocks of muscle

Gill slits

Sea Squirt Larva

Stiffening rod

Nerve cord

Great White Shark

In the jawed vertebrates, like this shark, the first gill arch became the jaws

Sea Squirt Adult

Water flow

In the sea squirts, only the larvae retain typical chordate features. The adults draw in water and filter feed, using a mesh-like 'bag' to trap food

Mesh-like bag

Gut

The eon of visible life is split into three major eras. The first of these, the Palaeozoic, or era of ancient life, began with the Cambrian period. But fast forward to around 400 million years ago (which would be around 30 November in our one-year timeline) we see that much has changed both on land and in the seas. This period, called the Devonian, is sometimes dubbed the Age of Fishes. It's a somewhat odd term, as fish have been successful in every era since their evolution, but the Devonian was home to an extraordinary array of different fish, while other marine vertebrates, like whales, dolphins or ichthyosaurs, had yet to evolve.

Swimming in Devonian seas would not be for the fainthearted. The hallmark of one extinct group of fish is the armour that covered their bodies, with the largest species achieving roughly the size of a great white shark. Called *Dunkleosteus* (pronounced: *Dunk-lee-ost-ee-us*), this powerful predator did not have teeth but was armed with a bony beak that featured sharp cutting edges and a cruel pointed tip. Opening extremely rapidly to suck in nearby prey, the beak could be snapped shut equally quickly, delivering a bite of extraordinary force to the unfortunate victim.

Gliding alongside these armoured giants were the ancestors of modern fish. Sharks patrolled the waters looking for easy prey, and while they closely resemble their modern descendants, we shouldn't fall into the trap of thinking that modern sharks are primitive creatures. All species continue to evolve, and today's sharks inevitably differ in many ways from their ancestors. Some of these differences are clearly visible – ancient sharks had their mouths at the end of their snouts, while in most modern sharks the mouth is set back from the snout, on the underside of the body. Other differences are harder to spot, because although the teeth of sharks fossilize exceptionally well, we can't examine the contents of ancient cells, long decayed.

Modern sharks are admired and feared by humans in equal measure. Our admiration no doubt stems from their sleek minimalist design, while our fear is certainly disproportionate, given the danger that they pose. Worldwide, there were a total of 57 unprovoked shark bites in 2020, of which ten were fatal, but every year in the US alone, around 40 people die from injuries related to dog bites. Most sharks confine themselves to eating fish and have no interest in humans, while a few of the largest sharks, like the whale shark, the basking shark and the deep-sea megamouth, have reverted to the filter-feeding ways of their ancestors. These ocean giants cruise through the water with their mouths permanently open, filtering out tiny organisms like krill or other small crustaceans.

Bony fish, although heavily outnumbered by their armoured relatives, were also found in Devonian seas. The familiar ray-finned fish that today sustain many human populations were generally small and rather uncommon, but lobe-finned fish were more abundant. While both types of fish sport two sets of paired fins on the underside of their bodies, the paired fins of lobe-finned fish contain thicker, more developed arm bones. And for a few lobe-finned fish living in shallow water during the Devonian, this arrangement allowed them to brace themselves against the bottom and raise their snouts into the air.

Given that oxygen dissolves in water, it might seem surprising that any fish would be interested in breathing air. But fish, like other vertebrates, are large, active animals that depend on muscle power. Muscles need lots of glucose and oxygen, and the size of vertebrate bodies creates problems in meeting their demands. In vertebrates, the greedy muscle cells are buried deep within the body, too far from the surface of the animal to get enough oxygen and too far from the gut to get enough food.

To deliver oxygen and glucose to the hard-working muscle cells, vertebrates have evolved a *circulatory system*. Kept moving by a muscular *heart*, life-giving blood travels around the body in a system of pipes, picking up and dropping off molecules in different places. To sustain their activities, the hearts of vertebrates beat continually throughout their lives, although the pace can change: hearts beat faster when the bodies they belong to are active, and slow down when their owners are relaxing. And it's not just the level of activity that matters – large animals have lower heart-rates than small ones. The heart of a blue whale beats around 35 times per minute, while a mouse manages an astonishing 600, but intriguingly, the hearts of most living vertebrates beat roughly one billion times before forever giving up (although a human heart can expect to beat twice as many times).

To extract vital oxygen from the water, all fish are equipped with *gills*. Formed from thousands of thin, feathery filaments bundled together and tucked away behind the eyes, gills are richly supplied with blood. The large vessels that bring the blood to the gills branch again and again, ending in thin-walled *capillaries* so narrow that red blood cells are forced to squeeze through one at a time, bringing each one into the closest possible contact with the surrounding water. Gases move between the blood and the water: oxygen diffuses in, while carbon dioxide diffuses out. To maximize this exchange, fish must maintain a flow of fresh water across the gills, which they create by actively pumping water through the gills or by continually swimming with their mouths open. Yet, even with this brilliant design, the supply of oxygen in water is limited, and fish struggle to supply their muscles with enough of this precious gas to keep them swimming.

Fish have two types of muscles: red and white, and only the red muscle is generously supplied with blood vessels. Fish use the red muscles for normal everyday swimming – pushing themselves forwards with a gentle flick of the fins or a more determined wag of the tail – and they can keep this up for a long time because oxygen is supplied to the muscles at the same rate that it's consumed. But, when startled or chased, fish can put on an impressive sudden burst of speed. The extra power comes from

The Structure of Gills

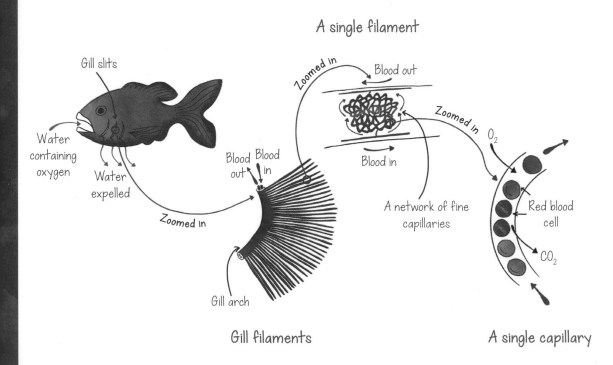

A single filament

Gill slits

Zoomed in

Blood out

Zoomed in

O_2

Water containing oxygen

Water expelled

Blood out Blood in

Blood in

A network of fine capillaries

Red blood cell

CO_2

Zoomed in

Gill arch

Gill filaments

A single capillary

engaging the white muscles, which have a limited blood supply. In the absence of oxygen, the white muscles resort to fermentation (anaerobic respiration), causing *lactic acid* to build up within two or three minutes, poisoning the muscles and forcing the fish to stop. So, if the fish is lucky enough to escape its pursuer, it will have to hole up, exhausted, for several hours, while oxygen trickles back into the white muscles and converts the lactic acid into a non-toxic molecule that can be usefully recycled.

During the Devonian, fish were common around the shallow seas that fringe the margins of the continents, swimming up estuaries and inlets, where rivers emptied their contents into the sea. The land adjacent to these coastal waters was colonized by the first forests and the water was probably full of rotting vegetation, a bit like mangrove forests today, providing food for bacteria and other organisms that cause decay. Their activities use up oxygen, leaving fish in danger of suffocation, so some evolved a simple *lung*, allowing them to supplement the meagre oxygen in the water with the plentiful oxygen in the air. Fossilized skeletons of these large animals have been recovered, revealing how one group of lobe-finned fish gradually acquired the means to spend more and more time out of the water. So much so, that they could eventually leave it altogether.

Adapting to land meant significant changes to the body plans of our tetrapod ancestors, including, for example, the evolution of lungs. It's easy to think that tetrapods are somehow better or more highly evolved than modern fish, but each organism is superbly adapted to its own environment, so we can't say that lungs are better than gills – as evidenced by the 320,000 people that the World Health Organization estimates die by drowning each year (which rather puts the ten shark-related deaths into perspective). Gills are a fantastic adaptation for extracting oxygen from water; indeed, they only fail when the water becomes so oxygen-depleted that there's rather little left to extract.

Similarly, we must remember that evolution can only take baby steps and crucially each step must offer an immediate advantage. Fish did not peer out of the water, see a land that they were desperate to take advantage of, and so miraculously evolve the required adaptations to do so. Instead, there must have been an immediate advantage to evolving a simple lung, which is why we think it probably evolved to allow its owner to gain extra oxygen in oxygen-depleted waters. We're confident that developing a simple lung can give a water-dwelling fish a real advantage partly because three species of *lungfish* can still be found on today's Earth – sporting gills and a lung – which allows them to survive when the shallow pools they usually inhabit dry up.

The ancestor of lungfish was the last modern fish group to split away from the shared ancestor of the tetrapods, making them the nearest living relatives of the tetrapods. The fossil record reveals how the tetrapod ancestor evolved a long snout, perhaps to snap food from the surface, and streamlined the number of fins – leaving only the two sets of paired fins under the body plus the tail fin. Gradually the single arm-bone that enters each of the paired fins thickened and became linked to the spine, allowing these fins to bear the weight of their owners. These changes meant that, by the close of the Devonian, large four-legged animals were paddling around the shallow waters, breathing air and snapping at fish, propelled by a long tail.

The early tetrapods were large animals, more than a metre in length, and as their limbs strengthened, they left the water to take advantage of the land. Perhaps the spur was freedom from the monstrous armoured fish that dwelt there, but although they could breathe air and walk around freely, the early land-dwelling tetrapods had to return to the water to lay their eggs, which developed into free-swimming tadpoles equipped with gills. Most of us associate tadpoles with frogs, toads or salamanders, which we collectively call *amphibians*, but we have to remember that modern amphibians might be quite different from their ancient ancestors.

Modern amphibians are much more specialized than the early tetrapods, but, they too remain tied to the water – requiring it to lay their soft, jelly-like eggs in and to provide a living environment for their tadpoles. Today's frogs are extraordinary animals, with moist, breathable skin that can be brilliantly coloured and patterned – and in the case of poison-dart frogs, produce enough venom to kill a large mammal. But, although large amphibian-like animals left the water and roamed the land during the Devonian, the reliance on water to rear their young prevented them from colonizing truly arid areas.

Stages in Tetrapod Evolution

Eusthenopteron: a lobe–finned fish

Tiktaalik: often dubbed the 'fishapod' because it's part–fish, part–tetrapod

Acanthostega: an early tetrapod.

Eventually, a new innovation freed the tetrapods from their aquatic past. Once the Devonian ended, the Earth began a new period, called the Carboniferous, in which towering forests covered the planet. During this time, one tetrapod evolved a new type of egg, enclosed in a flexible shell, that was more robust and resistant to drying out than the eggs of fish or amphibians. Within this protected environment, the fertilized egg developed directly into a juvenile – a small version of the adult – which could then break out and take its chances in the world. Many modern groups continue to produce such eggs: crocodiles, lizards, birds, turtles and even some unusual mammals, like the Australian *Platypus* and *Echidna*. However, this type of egg poses a challenge to groups like turtles, whose adults later returned to the water. Because the juvenile is no longer equipped with gills and could easily drown, female turtles must haul their heavy bodies out of the water to lay their eggs on the beach.

Alongside changes in their eggs, the adults of this new type of tetrapod evolved a scaly skin that reduced water loss, allowing them to move into more arid parts of the Earth. They almost certainly fed on insects, which had colonized the land ahead of the vertebrates, but they soon found other sources of food. By the end of the Palaeozoic, the world was filled with land-dwelling tetrapods of all shapes and sizes, exploiting every corner and feeding on plants, fish, invertebrates and even each other. Yet, despite their great diversity, every tetrapod – including us – shares the same basic body plan.

Compared to most invertebrates, tetrapods are large, active animals and face some unique challenges. Jellyfish can also be large, but crucially, jellyfish only contain two layers of cells – each just one cell thick – and the mass of a large jellyfish is just jelly. Jellyfish-jelly consists of water and secreted proteins, rather than packed layers of cells, so a jellyfish can make thick masses of jelly without worrying about how to provide that jelly with sustenance. Vertebrate bodies are different. They are not only large, but also many cells thick, so their bodies require a fundamentally different design. Cells on the inside are too far from the surrounding water or air for diffusion to supply their needs, so vertebrate bodies have evolved a system of *organs* to play specialist roles, allowing cells deep within their bodies to obtain all the things they need.

Like all vertebrates, tetrapod bodies are powered by muscle cells. But muscle cells are expensive to run: packed with mitochondria, they guzzle glucose and oxygen, and generate carbon dioxide as a waste product. Carbon dioxide dissolves in water, but it also reacts with water to form a weak acid. If too much carbon dioxide is produced, then the acidity (or *pH*) of the cell begins to change, and this affects the workings of the cell's enzymes. To prevent catastrophe, excess carbon dioxide is carried back to the lungs in the blood and excreted into the air. So, tetrapod lungs have a dual role: they replace oxygen and remove excess carbon dioxide.

Lungs aren't modified gills, but they perform the same function, allowing the exchange of gases between the blood and the environment. Only a small amount of oxygen dissolves in the watery part of the blood, so red blood cells act as specialist carriers. Each red blood cell must be brought as close to the air as possible, so they are forced through tiny thin-walled capillaries where they have about one second to pick up fresh oxygen before they return to the heart. To maximise the area for gas exchange, advanced tetrapod lungs are divided up into thousands of smaller air sacs – in humans, the internal area of the lungs is about the size of a tennis court

The evolution of lungs caused a change to the tetrapod circulatory system. To allow it to pick up enough oxygen, the blood that travels through the tiny capillaries of the lungs is slowed down as much as possible, and the loss of pressure leaves the blood sluggish. To regain momentum, the freshly oxygenated blood returns directly to the heart, where it is pumped out to the rest of the body in a separate circuit, carrying oxygen with it, and this is why we have two sides to our hearts: the right side pumps blood to the lungs and is less muscular than the left side, which pumps blood to the rest of the body.

During its passage around the body, the blood also travels to the cells lining the gut. Here, it can pick up small molecules, like glucose and amino acids: the digested remains of the food that the tetrapod has been eating. So, when the blood arrives at the muscle cells it is packed with oxygen from the lungs and food from the gut, which should mean that the muscle cells have everything they need. But, there are other organs inside tetrapod bodies aside from lungs and hearts.

The Basics of the Mammalian Circulatory System

The problem

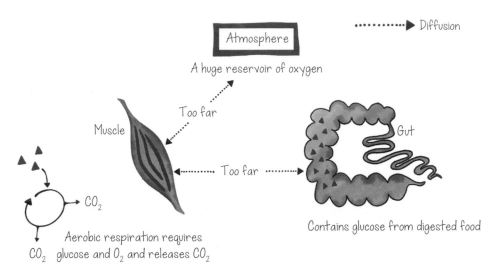

Atmosphere

A huge reservoir of oxygen

Diffusion

Muscle

Too far

Too far

Gut

CO_2

CO_2 Aerobic respiration requires glucose and O_2 and releases CO_2

Contains glucose from digested food

The solution

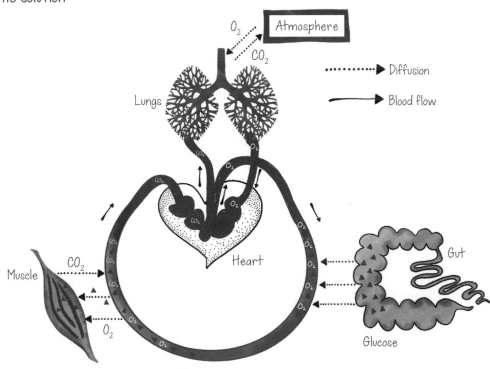

O_2 Atmosphere

CO_2

Diffusion

Blood flow

Lungs

Muscle

CO_2

O_2

Heart

Gut

Glucose

Animals can't run on sugar alone. They need to repair and replace their cells, and to do so requires other building blocks, such as amino acids to build new proteins and DNA letters to make copies of the all-important genome. The source of all these building blocks is the food they eat. When a lion eats a wildebeest, it digests the wildebeest proteins by breaking them down into individual amino acids. The cells that line the lion's gut then absorb these amino acids and the lion's blood delivers them around the lion's body where they are used to build lion proteins instead. But a lion is a committed carnivore, so it has lots of amino acids to spare, while it is short on glucose, which it needs to fuel its muscles if it's ever going to catch another wildebeest.

Fortunately, the lion doesn't have to worry because amino acids can easily be converted into glucose and used for aerobic respiration. But amino acids contain nitrogen, while glucose is a carbohydrate. The removal of nitrogen from amino acids creates ammonia (NH_3), which is toxic and must be quickly eliminated. For fish, this is relatively easy, because ammonia is highly soluble in water, so they can excrete it via their gills; but for tetrapods living on land, the removal of ammonia is much trickier.

In tetrapods, excess amino acids are sent to the *liver* for conversion. In mammals, the liver turns the nitrogen removed from the spare amino acids into a molecule called *urea*, which is soluble in water and much less toxic than ammonia, while in most other living tetrapods (birds, lizards, crocodiles and turtles) the excess nitrogen is turned into *uric acid*. Urea and uric acid are both excreted from tetrapod bodies by a paired organ called the *kidneys*, but the two molecules are excreted in slightly different ways.

In mammals, urea is excreted together with water in a liquid called *urine*. Urine is nothing more than highly filtered blood, and its formation begins when blood enters mammalian kidneys, and is forced to flow around tight knots of capillaries with leaky walls. The pressure forces the liquid part of the blood out of the capillaries and into the kidney's narrow tubes, while the blood cells and other large molecules stay behind. The kidney then retains the excess water and urea, passing the resultant liquid – urine – for storage in a waterproof bladder, where it can be excreted when convenient, and transfers all of the glucose and most of the water back to the blood.

The trouble with the mammalian system is that quite a lot of water is lost every day just getting rid of excess nitrogen – about 1.5 litres for an average human – but other living tetrapods are more efficient. Birds don't have a bladder, but instead send urine to an organ called a *cloaca*, where the uric acid precipitates out, allowing most of the water to be reabsorbed. The undigested food waste from the gut joins the uric acid in the cloaca and the two are despatched together as a soggy mixture of dark and light material, familiar to anyone who has been unfortunate enough to be pooed on by a bird (although it would be more correct to say that you were simultaneously pooed and weed on).

It's hard not to see this wasteful excretion of excess nitrogen as a fundamental flaw in animal biochemistry. Nitrogen is a precious resource and is often lacking from animal diets, particularly those that spend their lives eating plants. The poor wildebeest has to spend many more hours eating than the lion just to get enough amino acids from protein-poor grass; but for reasons unknown, amino acids can't be stored in animal bodies. Instead, the best thing that animals can do is to reclaim as much of the molecule as possible and excrete the resultant nitrogen-rich waste.

So, tetrapod bodies have evolved a set of organs to carry out specific functions, from oxygen delivery to waste excretion. The cells within these organs are highly specialized – under a microscope, a liver cell looks different from a kidney cell; indeed, two kidney cells from completely different species would probably look more similar than a kidney and a liver cell from the same animal. But all cells within an animal's body possess an identical genome, so how is it possible for cells within a single body to look so different?

The success of the highly organized tetrapod body plan, and indeed the body plans of many other animals, relies on exquisite regulation of gene expression. A liver cell is different from a kidney cell because it has turned on (or expressed) a different combination of genes to a kidney cell, otherwise it wouldn't carry out the right functions, and to do this it has to receive information that tells it, 'You are a liver cell'. This process begins right at the start of an animal's life; otherwise, organs might form in the wrong places or some organs might not develop at all. But how does a single fertilized egg cell tell every other cell in the developing body exactly what role it is expected to play?

Many of us have probably owned a Lego set at some point in our lives. Over the decades, Lego sets have become more sophisticated and there are now more than 2,000 different types of brick – available in a range of colours – and this allows people to create amazing sculptures. If you buy a Lego set – rather than loose bricks – then you already know what you're going to make (perhaps a dinosaur or a space station) and the set contains all the bricks that you need, plus an instruction manual that shows you how to fit them all together in the right order, so you won't be disappointed with the final result.

The bodies of complex organisms, like tetrapods, are also built from many different types of cell (or brick) – around 200 in a human body – and they too have to be assembled. But, if we compare the process of building a body to a Lego set, then the first thing to note is that we are going to be hugely disappointed when we open the box. All complex animals and plants begin life as a single fertilized egg cell – in Lego terms, just a single boring nondescript brick – and there's more disappointment to follow. The instruction manual is conspicuous by its absence, as it's been invisibly folded away inside the brick. So, it's safe to say that, at first sight, this appears to be the worst Lego set in the world.

On the upside, our disappointment is going to be short-lived. This living Lego set is self-assembling – and it's not just going to build itself, it's going to make each and every one of the millions or billions of bricks that it needs from the one and only brick that came with the box: all hail the fertilized egg cell!

As the fertilized egg cell divides, creating two daughter cells, it passes each one a copy of the genome – but the copies might already contain a few notes and annotations that have been put there by the egg cell. These alter the patterns of gene expression in the daughter cells – turning some genes on and other genes off – and this affects the way the cells look and how they function. When the daughter cells divide, they too pass on copies of the genome, to which further notes might be added, so gradually the daughter cells become more and more different, both to each other and to the fertilized egg cell from which the body originated. Finally, after many rounds of cell division, specialised cells that are clearly identifiable as muscle, nerve and blood cells appear.

A complex body in the early stages of development is called an *embryo*. As the cells within it continue to divide, they take on a range of shapes and sizes as they journey towards different fates. This process of *cell differentiation* is one of the key elements in the development of an embryo, and when it's finished, every cell in the body knows what kind of cell it is. But how do cells in the early stages of the embryo 'know' whether they should set their daughters on a path towards muscle cells, liver cells or kidney cells?

Fate determination is controlled by signals sent out by other cells, and the first signals might be sent by the all-powerful egg cell when it divides. If the genomes of each of the two daughter cells contain slightly different notes and annotations, then these may affect the signalling molecules that the daughter cells produce. As the embryo continues to grow, the signalling molecules affect the cells around them, causing them to switch on certain genes and switch off others; but cells further away don't receive these signals, so they switch on other genes instead. By releasing different signalling molecules in different places, all cells within the developing embryo receive a unique combination of signals that tells them *where* – and crucially – *what* they are, and this seals their fate.

For most cells, once their fate is finally determined, there's no going back. At an early stage of development, cells can be moved around inside an embryo without causing too many problems – they just receive new signals and change their behaviour – but once their fate is sealed, this ability is lost. In an adult body, liver cells are always liver cells and can only give rise to more liver cells – which is a problem if the liver is permanently damaged, because there's no way to regenerate liver cells from other cell types. Because the early cells of the embryo can give rise to nearly any type of cell, we call them *pluripotent*, while the fertilized egg cell is truly *totipotent* – meaning it has the ability to become *all* things. Recent technology has enabled cells with a specified fate to regain their pluripotency, and this has the potential to revolutionize medicine: if we know how to give these cells the right signals, they could be used to regenerate cell types in an adult body that have become damaged and can't be replaced.

The final aspect of embryo development is the shaping of weird and wonderful three-dimensional structures, like organs. Complex geometry requires co-ordinated

Principles of Animal Development

All new lives begin with a mighty fertilised egg cell

The fertilised egg cell rapidly divides to form a hollow ball of cells

Early in vertebrate development, a group of cells stream inside to form the main body cavity or gut

Insects

At an early stage in this fruit–fly embryo, a molecule is released by cells at one end of the body, which diffuses away slowly, setting up a gradient

This molecule lets cells know whether they are to form the front or back of the body

At a later stage of development, the body has multiple segments. Cells within each segment have switched on a different combination of genes

■ grow wings

▨ grow legs

These genes act like signals to make sure that the correct structures develop on each segment

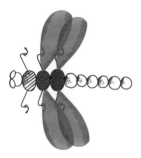

To save water, **desert** animals produce concentrated urine, and the **champion** is the **hopping** mouse

cell movements, and a particularly spectacular example occurs shortly after the fertilized egg cell begins to divide. After a few rounds of cell division, many embryos consist of a ball of cells with a hollow centre, but suddenly, at a single point on the outer surface of the ball, a group of cells begin to stream inside, as if a thumb was pushing into a soft hollow ball. Some of these cells are destined to form muscles and organs, and the tubular structure they create will form part of the gut.

The need for co-ordinated cell movements during development is one reason why bacterial cells have never been able to form complex three-dimensional bodies. Only eukaryotic cells can move with enough precision and co-ordination to sculpt sophisticated body parts, pulling themselves along using their dynamic cytoskeletons, or linking tightly together so that entire sheets of cells can be folded into intricate shapes, like the lining of the gut or the groove that will eventually form the spinal cord in vertebrates.

Once cell differentiation and the sculpting of body parts is complete, the embryo simply has to get bigger — although not all parts will grow at the same rate. For birds, lizards, crocodiles and turtles, this growth takes place within the protective environment of the egg, and some mammals do the same. Today, two types of *monotreme* mammals, the platypus and the echidnas, still lay eggs. *Marsupial* mammals, like kangaroos and koalas, give birth to a tiny and only partially developed baby, which crawls into a pouch to finish growing by feeding on its mother's milk. The *placental* mammals, to which most modern species of mammal belong, retain the embryo inside their bodies for longer, and the mother transfers all the building blocks that the embryo needs via a specialized organ called the *placenta*.

The fact that a few mammals share the same type of egg as lizards and birds reveals their shared ancestry. But birds, lizards, crocodiles and turtles are more closely related to each other than they are to mammals because the shared common ancestor of mammals split away first from the shared ancestor of other surviving land-living vertebrate groups. To find out how this happened, we need to return to our timeline, just a few million years after the Palaeozoic ended and the Mesozoic began.

As the Mesozoic dawns in what is now South Africa, a one metre-long, scaly tetrapod with a massive head emerges from a burrow and rootles in the earth with the help of two tiny tusks. Scissoring off a tough plant stem with its sharp teeth, it stares at a neighbour who dares to move too close and grunts a warning, which echoes off the surrounding hills. Scanning around, a casual observer might note that the entire area is filled with other individuals of the same species – indeed, if we could travel around the supercontinent of Pangaea at this particular point in time, we would be continually confronted with the same face.

Lystrosaurus (pronounced: *List-roh-saw-us*) is a great survivor – one of the few large animals to make it through the Great Dying: the mass extinction that wiped out most of its competitors and predators at the end of the Palaeozoic. The Great Dying was caused by sudden sustained volcanic activity that poured out carbon dioxide and hiked up global temperatures. It was the biggest of all mass extinctions and within a very short time span over 95% of life in the sea was lost – and some groups, like the trilobites, were never seen again.

It's hard to say why *Lystrosaurus* survived when others didn't: it could burrow, which might have given it respite from the ever-increasing heat, and it wasn't a fussy eater, so perhaps it could keep finding food as the plants around it changed. But whether by design or through some cosmic whim, for the few million years that followed the Great Dying, *Lystrosaurus* had the world practically to itself – in parts of what is now South Africa, nine out of 10 fossilized skeletons from this period belong to this creature – but it wasn't entirely alone.

Lying alongside *Lystrosaurus* in the bone beds of South Africa are the fossilized remains of a few other large animals. One of these, called *Proterosuchus* (pronounced: *pro-tare-oh-sue-cuss*), was more than three metres long, looked a lot like a modern crocodile and almost certainly preyed on *Lystrosaurus*. But while we might want to give the name *reptile* to both of these animals, they actually belong to two very different lineages.

During the Palaeozoic, another great split had emerged within the land-living tetrapods. Although rather similar in many ways, one group went on to found the mammals, while the other eventually gave rise to a whole suite of modern vertebrates including lizards, turtles, crocodiles and birds, as well as many extinct groups like dinosaurs, icthyosaurs and pterosaurs. *Lystrosaurus* belongs to the lineage that eventually gave rise to mammals (and could be your ancestor), while *Proterosuchus* belongs to the sister lineage that gave rise to all the other creatures that we might want to call reptiles (and can't be your ancestor unless you are a bird with extraordinary talents and have managed to read this book). This creates something of a problem. Do we want to call *all* land-living tetrapods (apart from amphibians) reptiles? Or do we want to reserve that word for the lineage that gave rise to modern lizards, turtles, crocodiles and birds, as well as other extinct groups, like pterosaurs, ichthyosaurs and dinosaurs?

Most current experts prefers to use the word *reptile* to mean the latter. This means that birds are reptiles, and although this might seem odd, it's absolutely clear that birds are the direct descendants of a group of predatory dinosaurs that walked on two legs. They were called the *therapods* and include animals like *Velociraptor*, who played a starring, albeit not strictly accurate, role in Jurassic Park. We also have something of a problem when we refer to an animal like *Lystrosaurus*. This creature certainly belongs to the lineage that gave rise to mammals, but we can't call it a mammal because the key features of mammals, like hair, had yet to evolve, but we can't really call it a reptile either (strictly speaking it is a *synapsid*).

Dinosaurs are highly specialized reptiles that were the dominant group of land animals for around 170 million years. One of the (many) reasons that humans love dinosaurs is that we live in a world filled with mammals and birds, and it's incredible to imagine a world dominated by scaly-skinned giants. Mammals and birds are typically rather small animals – with many weighing less than a standard bag of sugar – but the defining feature of dinosaurs is their size. No known species was smaller than a large rabbit and some weighed a frankly unimaginable 70 tonnes (about the same mass as the Space Shuttle).

Dinosaurs arose from the impoverished world of the early Mesozoic, where a few omnipresent beasts held sway. The first period of the Mesozoic is called the Triassic, and by the time it ended, the greatest reptiles to walk the Earth had truly begun to stamp themselves on the face of our planet – and during the Jurassic and

Cretaceous periods that followed, no visiting alien could have been in any doubt about who was in charge. So why and how were some dinosaurs so enormous?

In films, changing the size of an animal or a human is simply a matter of applying a magic ray gun seized from a malevolent villain, but the reality is somewhat different. We have already seen how the evolution of vertebrates required fundamental changes to their body plans, including the development of multiple organs, like lungs and hearts. But given a specific body plan – like that of a mammal – is it feasible to simply apply the magic ray gun and create a mouse the size of a house or an ape the size of a grape?

To answer this question, let's consider a very simple shape, like a cube. If we start with a cube whose sides measure 1 cm, then its volume is 1 cm^3 (1 x 1 x 1) and the surface area of a single face is 1 cm^2 (1 x 1). If we now apply the magical zapping ray to the cube, we can enlarge it, so that each side now measures 2 cm. Surprisingly, we now find that the volume of the cube has grown to 8 cm^3 (2 x 2 x 2) while the surface area of a single face is 4 cm^2 (2 x 2). These seemingly innocent numbers are worrying. When we zapped it, we only *doubled* the length of the side of the cube, but we increased the surface area *four* times and the volume *eight* times. So, although the two cubes are exactly the same shape, the larger one has a much smaller *surface area to volume ratio* than the smaller one. And this ratio is absolutely crucial for just about every aspect of an animal's biology.

As we saw from zapping the cube, when animals get larger, their volume increases faster than their surface area, and this imbalance causes all sort of problems. One critical factor for any mobile animal is the relationship between power and weight. Weight is proportional to volume, so the larger the animal, the more it weighs; but the power to move that weight comes from an animal's muscles. The strength of muscles isn't proportional to their volume or their mass, but rather to their area in cross-section – so the power of the biceps muscle in a person's upper arm is determined by how thick it is, and not by how heavy it is. The upshot of this unfortunate fact is that large animals are seriously under-powered, while small ones appear to be almost bionic.

Body Size Scaling

1cm	Radius	4cm
12cm²	Surface	112cm²
4g	Mass	113g
¹²/₄=3	Surface/Mass	¹¹³/₁₁₃=1

A mouse is like a small sphere, with a large surface for its mass

An elephant is like a large sphere, with a small surface for its mass

The mouse loses heat rapidly

The elepant loses heat slowly

So, a 30g mouse must eat its own body weight in food every 6 days

An elephant only needs to eat its own body weight in food every 30 days

A
leaf-cutter
ant can
lift nine
times
its
own
body mass

The effect of the declining power-to-weight ratio can easily be observed by watching animals of different sizes. Mice can leap into the air from a standing start, while elephants are constrained to swing one leg ponderously after another; small birds, like sparrows, hop about with two feet together, while larger birds, like eagles, are forced to walk – and these differences are all due to size. The power-to-weight ratio declines as animals get bigger and this places an upper limit on how big an animal could ever be – eventually they would reach a size where they would be so heavy that their muscles simply couldn't move their massive bulk around.

The power-to-weight ratio isn't the only feature of an animal's biology affected by its size. All tetrapods generate some heat internally through aerobic respiration, but some, like modern lizards, warm themselves largely from the environment, by basking in the sun or draping themselves over hot rocks. This stratagem is called *ectothermy* (pronounced: *ect-oh-therm-ee*) and it's cheap to run, but leaves a lizard sluggish during the night or during the winter. The alternative, practised by modern birds and mammals, is *endothermy* (pronounced: *end-oh-therm-ee*), where most of the heat is generated internally by cranking up the metabolic rate and burning sugar day and night to keep the temperature inside stable, whatever the weather (ectothermy and endothermy used to be referred to as cold- and warm-blooded, but these terms are no longer used by biologists).

These strategies have implications for dress code. A lizard dares to bare, so that it can absorb the heat of the sun, while a mouse needs to cover itself in fur to stop internally generated body heat from escaping. But whether ectotherm or endotherm, size has critical implications for the lifestyle that animals can pursue.

As tetrapods get larger, they inevitably have more cells, and each cell generates a small amount of heat. Heat is lost through the surface of an animal's body, so as an animal gets larger – and the surface area-to-volume ratio declines – it becomes increasingly difficult to lose the heat that their cells generate. For endotherms, this is a great advantage. Mammals and birds maintain a high body temperature, so their cells must have a high metabolic rate, which is costly and requires eating large amounts of food. But, crucially, maintaining a high body temperature gets easier as mammals and birds get larger. In fact, today's largest mammals, like elephants and rhinos, are so large that they are in danger of overheating and don't require a fur coat.

Indeed, large animals retain heat so much better than small ones that they can reduce the metabolic rate of their cells. This means that they actually have to eat a lot less per unit body mass. This allows them to eat low-quality food, like grass, or the leaves of trees, that don't contain much nutrition. It's simply impossible for an animal the size of a mouse to survive on grass as it would need to eat for more than 24 hours per day. Instead, small mammals have to specialize in energy-dense food, like insects and seeds, and even so, they are always in danger of starvation. The shrew's astonishing metabolism – which is required to maintain its body temperature – means that it has to eat 80% to 90% of its own body weight every day. So, shrews are about as small as a mammal can possibly be.

Ectotherms are the other way round: a truly enormous modern lizard would struggle to absorb enough heat from the environment to get itself moving in the morning, while a small one can be quickly up and running. Which rather brings us back to the question of how dinosaurs – which are also reptiles – became so unfeasibly large.

Although all dinosaur groups had giants, the true titans belong to a single group – the *sauropods* (pronounced: *saw-roh-pods*). These monsters supersized in the middle of the Jurassic and remained the dominant large herbivores until the mass extinction at the end of the Cretaceous brought their reign to a close – a time span of around 120 million years. Their basic shape never really altered: all sauropods have four large columnar legs supporting a massive body with a long neck and equally long tail – on the lines of the well-known favourite, *Diplodocus*.

It's hard to comprehend the size of sauropods, as we have nothing remotely like them on today's Earth. An elephant is certainly a large animal, but it would be utterly dwarfed by a sauropod, and only if a person were to sit on top of a typical three-storey house, could they look most sauropods in the eye (leading some of us to wish fervently that they would one day parade down the street).

It's important to ask not just *how* these creatures grew so large, but *why* they

The Relative Size of a *Brachiosaurus* and an African Elephant

did. Large size appears to have many benefits, such as the ability to eat low-quality food, but another advantage is that an extremely large animal is extremely difficult for a predator to kill. In support of this idea, the largest herbivores on Earth today are generally larger than the largest carnivores: elephants and rhinos are much larger than lions, and they don't fear them too much either. So, adult sauropods (although not their babies or juveniles) would have been largely immune to even the most fearsome carnivores, such as the therapod dinosaurs, which, in addition to *Velociraptor*, included *Tyrannosaurus rex*.

One downside of being gigantic is that it requires eating an enormous amount of food. Although large animals need *less* food per unit body mass, they still need more in total. Sauropods were herbivores and managed to eat the necessary amount of plant material every day by doing away with any kind of chewing. They simple stripped away leaves and branches with their small teeth and swallowed them whole – and they could access food from all levels by standing still and swinging their long necks around in a wide arc. This highly efficient way of feeding was only possible because the head of a sauropod was tiny and lightweight when compared to an elephant or rhino. Indeed, sauropod heads seem almost ludicrously small, although their brain size is pretty typical for a reptile of this size (lack of brain power being no barrier to world domination). But even assuming that they could get enough food, a long-standing question remains about how dinosaurs maintained their body temperatures. Did they primarily gain heat from the environment, like a modern lizard, or did they generate enough heat internally to keep a constant body temperature, like a mammal?

Current evidence suggests that sauropods, like other dinosaurs, must have generated a lot of their heat internally and so had higher metabolic rates than modern lizards. High metabolic rates are needed for rapid growth, and this was necessary because sauropods started life in an egg that was only slightly bigger than those produced by an ostrich. The baby sauropod, perhaps weighing around 1.5 kg, would then have had to increase its mass by a factor of at least 10,000 to reach adult body weight, and as this took decades rather than centuries, it must have been fuelled by a decent metabolism. This might lead us to think that the adults would surely have overheated, but the metabolic rate could have declined as the sauropod grew, and the long thin neck and tail would have helped to dissipate excess heat, rather like the huge ears of an elephant.

Whether or not dinosaurs were endothermic has always been controversial, but there is now strong evidence to support it. Mammals aren't the only endothermic tetrapods on our planet today: birds maintain a body temperature of around 40°C, and they evolved from the enemies of the sauropods – a group of therapod dinosaurs. This astonishing claim was only taken seriously in the 1970s, although Thomas Henry Huxley, a friend of Darwin's, had first suggested the link in the nineteenth century. The evidence for birds being a type of dinosaur is now overwhelming, and many small feathered therapods have been retrieved from early Cretaceous rocks in China. So, while a herd of sauropods may never visit a road near you, dinosaurs are – thankfully – still with us.

The end of the great sauropods and most other types of dinosaurs, occurred when an asteroid collided with Earth around 65 million years ago, ushering in the Cenozoic era. But, of course, this was not the end of the tetrapods. Mammals and birds made it through this apocalyptic time – as did crocodiles, turtles and lizards – and they diversified into thousands of new species, some of which still populate our planet.

All modern tetrapods owe their existence to their fishy ancestors that lived in oxygen-depleted waters and evolved a primitive lung to help them survive there. But that ancestor would never have left the water if there wasn't something on dry land to attract them. During the Cambrian, the continents would have been barren and mostly devoid of life, but gradually, some new pioneers found a way to colonize the land and cover it in a cloak of green. These pioneers are the land plants, who, armed with some stolen technology, took over the continents and provided a home for the land-living tetrapods.

Chapter 9
PLANTS
Truly, madly, deeply green

The fossil record of the first tetrapods is highly revealing, allowing us to follow their trail out of the water around 370 million years ago, during the late Devonian. The bony skeletons of vertebrates mean that they are often caught in the headlights of the fossil record, but many soft-bodied groups, like the jellyfish, have ghosted between the Cambrian and the present day without leaving so much as a trace. Catching a glimpse of the past lives of the more elusive inhabitants of the Earth requires exceptional events or conditions, and just occasionally, these glimpses happily coincide with a pivotal moment in the history of our planet.

Near the small Scottish village of Rhynie, an exceptional fossil site has yielded remarkable clues as to how one group of organisms transformed the continents from their original barren state. In the early Devonian – 410 million years ago – the site was occupied by bubbling hot springs, rather like those found today in Yellowstone National Park in the USA. Hot springs rely on volcanic activity, and their silica-rich waters can sometimes bubble over, entombing their surroundings in a glassy rock called a chert. Entombed within the Rhynie cherts are the flattened remains of simple branching structures that were pushing their way into the ancient air before they were abruptly boiled alive. These are some of the earliest *plants* and we know exactly what they looked like because their bodies have been almost perfectly preserved.

The Rhynie cherts reveal intimate details of the first pioneering greenery, including individual cells and their contents. Standing only around 20 centimetres tall, these bold adventurers were the first true colonisers of the land, and they set the stage for the tetrapods that followed. During the Cambrian, when life exploded in the oceans, the land was still a hostile place – bare rock with little or no soil. Plants eventually succeeded against the odds because they were carrying stolen technology, which allowed them to photosynthesize and conjure sugar out of thin air. Armed with this valuable resource, they then bartered with organisms like fungi to obtain other essential goods – like scarce mineral ions that are hard to find on land. Their success transformed the continents and the sky above our heads. Plants create soils, influence weather patterns and provide food for hungry animals. Indeed, plants continue to offer extraordinary services to all life on our planet, including ourselves, to this day.

Plants evolved from *algae* that lived in ponds and streams. All plants share an

Early Land Plants From the Rhynie Cherts

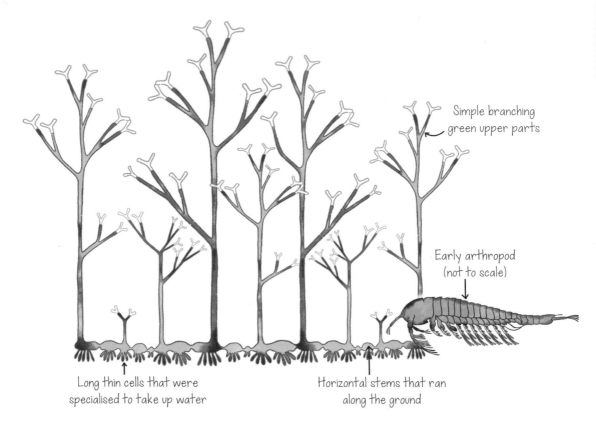

Simple branching green upper parts

Early arthropod (not to scale)

Long thin cells that were specialised to take up water

Horizontal stems that ran along the ground

exclusive common ancestor, so only one algal species successfully made the transition from water baby to landlubber. The challenges faced by algae when moving out of the water and into the air are immense, so perhaps it's unsurprising that all plants can be traced back to that first pioneer. We don't know if the algae in question were already multicellular before they ventured out of the water, but living on land certainly required a body in which specialized cells could work together to solve the problems presented by their new environment. But let's rewind for a moment and consider where the algae that would eventually give rise to the land plants came from.

Algae – like plants – make sugar through photosynthesis. Sugar is a carbohydrate and so requires only three elements: carbon, oxygen and hydrogen. The first two are contained within carbon dioxide (CO_2), which dissolves freely in water, but unfortunately hydrogen is much more difficult to obtain. Most of the hydrogen on Earth today is held within the grip of oxygen atoms inside molecules of water (H_2O), and the oxygen atoms are highly reluctant to release it.

Around 2.5 to 3 billion years ago, cyanobacteria cracked the central difficulty of photosynthesis by evolving an incredible molecular machine, called a photosystem. The machine harnesses the sun's energy and uses it to smash up water molecules, so freeing the hydrogen within. A far more disruptive invention than the wheel or the iPhone, the photosystem changed the biosphere forever, as each one releases unwanted oxygen from the shredded water molecules back into the atmosphere. As oxygen levels increased, many bacteria met their doom, but others took advantage of this strange new gas by evolving aerobic respiration – a powerful way to extract energy from carbohydrates – which allowed them to recharge truckloads of ATP batteries, which led to a supercharged metabolism.

Next, an unexpected alliance – between an archaeal cell and a bacterial cell that could carry out aerobic respiration – gave rise to the eukaryotes, who converted the enslaved bacteria into organelles called mitochondria. Now powered by aerobic respiration, eukaryotes made a successful living by engulfing other cells in order to obtain fuel and essential building blocks. But, if they could carry out photosynthesis, then they could make their own fuel and building blocks from little more than air and water. Surely, the wily eukaryotes wouldn't let such a great trick pass them by?

Evolving a process like photosynthesis is much more difficult than evolving longer legs or wings of a different colour. Photosynthesis is fiendishly complicated, requiring delicate machinery and a host of new enzymes. The cyanobacteria evolved photosynthesis in stages, beginning with a simpler photosystem that used the sun's energy to remove hydrogen from hydrogen sulphide (H_2S) and only later moving on to smashing up water (H_2O). But the eukaryotes didn't have to reinvent the wheel. They had already stolen one piece of valuable technology when they enslaved mitochondria, and because they regularly engulf other cells, they had the means to do it again. Eukaryotes are a diverse bunch, but around 1.5 billion years ago, the first *alga* engulfed and enslaved a cyanobacterium. There is a certain poetry in this event, as by stealing the technology for photosynthesis, the eukaryotes had appropriated the invention that poured oxygen into the Earth's atmosphere, and so spurred their evolution in the first place.

The first algal cell to carry an enslaved cyanobacteria must have had a huge advantage over its competitors, as it could make its own food and no longer needed to hunt out and digest other cells. Over time, the captured cyanobacteria and its host became tightly interwoven and today, we consider the enslaved cyanobacteria inside algal and plant cells to be just another organelle, called a *chloroplast*. Like mitochondria, chloroplasts can't leave their hosts or live independently, as most of their original genome has been transferred to the nucleus of the host cell. But by teaming up with a eukaryote, cyanobacteria eventually conquered the land as well as the oceans, making them hot contenders for the most successful organism of all time.

One common misunderstanding is to imagine that plant cells contain chloroplasts *instead of* mitochondria. But nearly all eukaryotes contain mitochondria, including plant cells. Mitochondria are necessary to recharge ATP batteries and keep the plant cell running during the hours of darkness when the photosystems are inactive. Indeed, when it's dark, plant cells only conduct respiration, and not photosynthesis, so they actually release carbon dioxide and take up oxygen – although in much smaller amounts than active animals like us. In the past, this concern led over-zealous nursing staff to remove flowers from the bedsides of patients at night, fearing suffocation, but the amounts of oxygen that plants consume wouldn't pose a risk to a human, even if their bedroom was packed to the rafters with pot plants.

The first true algae were single-celled, but single cells can easily be swept to the bottom of the ocean, away from the sunlight on which they depend. To keep themselves anchored near the surface, many algae form multicellular bodies – called seaweeds – that can be seen on a trip to the seaside pretty much anywhere around the world. Some are red, some are brown and some are green, and each of these is a different kind of alga that independently evolved the multicellular habit. Only the green and red forms are true algae – direct descendants of the original eukaryotic cell that captured the cyanobacteria. The brown forms, like kelp, probably evolved later by capturing and enslaving a single-celled true alga, rather than a cyanobacterium.

Multicellular red and green seaweeds consist of a thin sheet of algal cells, but kelps are more robust, forming complex multicellular bodies. Growing up to ten metres in length, a kelp is tethered to the sea floor by a small *holdfast* that supports a single stalk that branches out into one or more thin rubbery blades. Growing in temperate waters, kelps can achieve astonishing growth rates, putting on half a metre per day when conditions are ideal. Often destroyed by wave action during the winter, new individuals must regrow their tangle of long brown fronds each year, and within this protected environment, kelp forests support a spectacular diversity of snails, jellyfish, sea urchins and marine animals, as well as forming vital nursery grounds for commercially important fish.

Just like an animal, all seaweeds begin their lives as a single cell, which gives rise to a body in which all cells share the same genome. The cells secrete walls that provide support, but most have no other kind of skeleton, so their flat bodies flop around hopelessly when taken out of the water. Their bodies are flat rather than round for the rather obvious reason that all cells in a seaweed photosynthesize and there's no point having cells buried deep within a body where sunlight can't reach them. But a seaweed can't live on carbohydrates alone – among other elements it needs a source of nitrogen to make proteins and phosphorus to make DNA. Luckily for seaweeds, seawater is rich in minerals and the cells in their thin flat bodies can directly import all the molecules they need from the water in which they are bathed. So, with the exception of the much larger brown algae, it's perhaps unsurprising that seaweed bodies aren't very sophisticated and consist of rather few different types of cell.

Many seaweeds live in the tidal zone, where twice a day the sea retreats and they are left out of their element. To prevent themselves from turning into seaweed crisps, many secrete slimy substances, called *alginates*, that reduce water loss, and beneath this protective layer they simply sit things out until the water returns and they can photosynthesize again. Humans have found many uses for alginates, and kelp is harvested around the world, finding its way into food and medicines. It is used to thicken many processed foods and forms the perfect base for gels that can be applied to human wounds, providing a moist anti-bacterial environment in which skin can repair itself.

A large tree can suck a tonne of water out of the soil in a single day

Although undoubtedly a useful adaptation, alginates will not protect seaweed cells forever once they are exposed to the air. So, when algae moved permanently into the sun, they needed to make much more serious changes to their bodies.

The biggest problem for an aquatic organism when it moves onto land is how to prevent catastrophic loss of water. To solve this problem, the early land plants evolved a *cuticle* – an outer layer of cells that secretes a fatty waterproof layer. This seems like a great idea, but it immediately creates a second problem – if water can't pass easily through the cuticle, then nor can air, and air contains carbon dioxide, which is fundamental for photosynthesis.

Land plants solved this problem in an ingenious way by peppering the cuticle with tiny holes, called *stomata*, through which carbon dioxide could enter and oxygen could leave. Water is a small molecule, and also passes through the holes, but each hole is surrounded by two specialized *guard cells* that either push apart to open the hole or squash together to close it, and this gives the plant control over the loss of water. If water is in short supply, then the plant can close the stomata and reduce water loss – although this comes at a price. If the stomata are closed, then carbon dioxide can't enter, so a plant without water is paradoxically in danger of starvation. We know that stomata were an early innovation because amazingly we can see the guard cells quite clearly on the fossilized plants in the Rhynie cherts.

Closely connected to the problem of water loss is the problem of how to obtain water in the first place. In many places on the planet, it rains quite a lot, so it's easy to imagine that water is plentiful on land. But the air normally contains very little water – whereas plant cells are full of it. The second law of thermodynamics doesn't allow differences like this to persist, so water molecules continually diffuse out of plant cells through the stomata to the surrounding air – in exactly the same way that water molecules move out of wet clothes when they are hung on the washing line. But, unlike wet clothes, if a plant can't replace the lost water molecules, then its cells will collapse and the plant will die. So, it's absolutely essential that plants replenish the water lost to the air by drawing new water molecules from somewhere else.

On land, there is much more water in the ground than in the air. When it rains, water collects on the soil surface, and soils are pretty good at holding water in the tiny spaces between their particles. The early land plants didn't have proper roots, but some of their stems ran horizontally along the ground, and these stems were covered in a profusion of what appear to be tiny hairs. Closer inspection reveals that each hair is a long thin cell that pushed down just a few millimetres into the soil and was specialized to take up water. Once inside the plant, the water taken up from the soil has to be moved to the leaves, which are in constant danger of being sucked dry by the air around them. But water is heavy and moving anything upwards means going against gravity.

The movement of water through a plant from the roots to the leaves is only possible because water is a rather remarkable molecule. Within each molecule of water, the single oxygen atom pulls the electrons that it's supposed to be sharing with the two hydrogen atoms towards itself, giving it a slight negative charge and leaving the two hydrogen atoms with a slight positive charge. As we have already seen, this means that water molecules are attracted to each other, making water 'sticky' – and this stickiness makes swimming difficult for tiny cells, like bacteria. But the same property works to the plant's advantage because the stickiness holds the thin column of water together, allowing it to be raised high above the ground.

In modern flowering plants the structures that transport water from the soil to the leaves are made from tissue called xylem (pronounced: *zy-lem*). Xylem is constructed from living cells that join together end to end and secrete a thick cell wall, but then they die, leaving a narrow pipe that water molecules can easily cling to. Water enters the bottom of the pipes in the roots and forms a continuous column through the plant's stem all the way up to the leaves, where a branching network of smaller pipes distributes water to the cells within each leaf. The water molecules delivered by the xylem replace those lost to the air and, as they move out of the top of the pipe, they drag the whole column of water onwards and upwards against gravity.

The total amount of water lifted from the ground to the air by the world's plants every day is immense. A large tree in full leaf can easily pipe up a tonne of water in a single day, and most of this will exit through the stomata and be released into the

How Water Moves Through a Plant

A Whole Plant

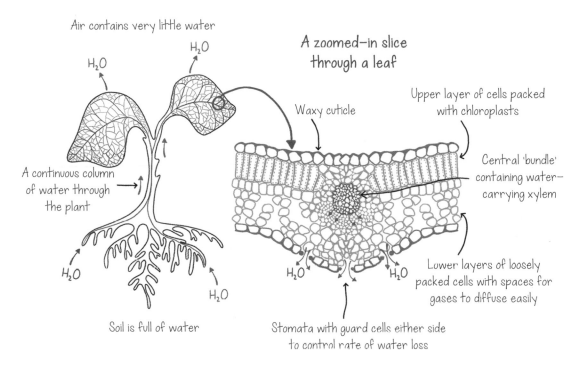

Air contains very little water

H_2O

H_2O

A zoomed-in slice through a leaf

Waxy cuticle

Upper layer of cells packed with chloroplasts

A continuous column of water through the plant

Central 'bundle' containing water-carrying xylem

H_2O

H_2O

H_2O

H_2O

Lower layers of loosely packed cells with spaces for gases to diffuse easily

Soil is full of water

Stomata with guard cells either side to control rate of water loss

air where it helps to form clouds. In a rainforest, enough water gathers in the clouds during the day to trigger heavy rain by late afternoon, which soaks into the thin tropical soils and is immediately piped back up again – and this recycling of water is uniquely efficient in forests. If they are removed or replaced with other vegetation, then rainfall will decline, pushing regions into drought and preventing forests from regrowing.

Water is needed to support the plant as well as provide the hydrogen for photosynthesis. Plants lack a hard skeleton and are in danger of flopping like a seaweed, so they rely on water pressure to help them defy gravity. The cells of plants secrete a box-like wall, made from *cellulose* – a long branching molecule that is relatively cheap for the plant to make – but the cell walls aren't strong enough to support the plant alone.

To see how water pressure helps the plant to stay upright, imagine that we were given the frustrating task of building a tall tower using some rather flimsy cardboard boxes. Every time we add new boxes to the top of the stack, the ones at the bottom collapse because their walls aren't strong enough to take the weight, and we begin to think that the task is impossible. Now imagine that we are also provided with some balloons. If we blow up a balloon inside each box, so that it pushes against the inside walls, then it will strengthen the walls of the box, allowing us to stack them much more easily – and we should find that we can now complete the task (and perhaps feel rather smug).

Plant cells resemble thin cardboard boxes with balloons inside. The balloon represents the plant cell, filled with water rather than air, and it secretes a wall around itself, the box, to stop the cell from bursting as it sucks up more and more water. The force exerted on the walls by the bulging cell is called *turgor pressure* and it prevents stems and leaves from wilting, as long as water is plentiful. You can easily see and feel the difference that turgor pressure makes by leaving a pot plant for a week or so without water – although be careful not to leave it too long or it might suffer permanent harm.

To maintain turgor pressure, plant cells must keep drawing water inwards, and this happens because of a process called *osmosis*. Osmosis is the movement of water across a membrane from a dilute solution to a more concentrated one, and is just a special case of diffusion. If two adjacent cells contain different concentrations of a dissolved molecule, like sugar or salt, then water will move until the concentration inside the two cells is the same. This is simply the second law of thermodynamics in action: it's the water that moves, rather than the dissolved sugar or salt molecules, because they can't cross cell membranes freely, while water can normally travel unhindered. So, as water from the soil exits the xylem vessels, it soaks into the cell walls like blotting paper, and is drawn through the membranes into the sugar-filled cells by osmosis, until the cell walls are stretched so tight that they prevent any more water from entering.

Turgor pressure is essential for plants, but it will only ever take them so far. To pipe water one hundred metres up into the sky, plants need to reinforce the xylem with tough fibres (like those found in a stick of celery) or, even better, with wood. Wood is made from a molecule called *lignin*, which is remarkably difficult for other

Turgor Pressure Keeps Plants Upright

OSMOSIS: water will move across a cell membrane when the concentration of small molecules (like sugars or salts) is different

The cells in your body are bathed in plasma, which contains the same amount of sugar and salts as the cells

Dissolved sugars and salts

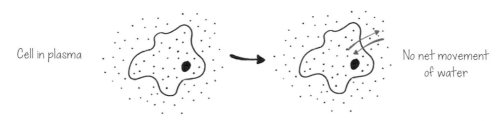

Cell in plasma

No net movement of water

But plant cells are bathed in rainwater drawn up from the soil. Rainwater contains very few small molecules, so an animal cell bathed in rainwater would burst

Cell in rainwater

Water moves in and cell bursts

Plant cells are surrounded by a cell wall

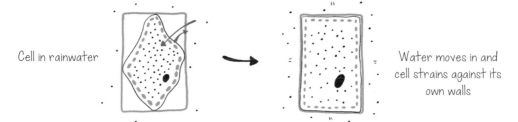

Cell in rainwater

Water moves in and cell strains against its own walls

TURGOR PRESSURE: the pressure exerted by swollen plant cells as they push against the cell wall

Not enough water – the plant wilts

Plenty of water – the plant is turgid

organisms to digest, making it extremely unappealing to hungry animals and therefore very long-lasting. As a consequence, trees are some of the oldest living things on our planet, with typical lifespans of several hundred years, and a record-breaking five thousand years for the gnarled and twisted bristlecone pines that inhabit the White Mountains of California. But, despite their longevity, the heart of every tree is dead – a mass of old xylem that long since ceased to function.

Every year, a tree adds girth to its ever-widening trunk by building brand-new xylem just beneath the bark. New pipes are necessary because old ones eventually stop working and have to be replaced. The pipes are most at risk during hot weather, when the soil dries out, making it harder and harder for the roots to extract water. As the sun beats down on the leaves, water molecules are drawn out of the top of the pipes, but the plant struggles to extract replacement molecules at the bottom. Eventually, the strain on the column of water is too great and the bonds between the water molecules break, leaving an air bubble in the pipe. A small air bubble can sometimes be bypassed, but when they are beyond repair, the tree simply seals them off. The heart of a tree is entirely made from these old plugged-up xylem and the living wood is just the thin outer layer beneath the bark – one of many reasons why people really shouldn't carve their names into it.

In a seasonal climate, each new set of pipes forms a visible ring in the trunk. These rings can be counted to reveal the age of the tree, while the width of the rings also carries useful information about the conditions for growth. In mild years with plentiful rain, the tree makes a wide ring filled with new pipes, but if it's too dry or too cold, then the ring is correspondingly thin. The link between the width of the ring and the weather allows us to reconstruct past climates, and sometimes reveals catastrophic events. In 1783, the Laki volcano in Iceland erupted, spreading dust into the upper atmosphere around the northern hemisphere and causing temperatures to plummet. Records obtained from white spruce trees in Alaska show that the growth ring in 1783 is tiny, and the local Inuit tribes still maintain an oral memory of 'the time summer did not come'.

Water is essential for the growth of a tree to power photosynthesis and maintain turgor pressure. But although carbohydrates can be made from water and carbon dioxide, the plant also needs to make new proteins and DNA, and to do that it needs to find other essential minerals, like nitrogen and phosphorus. Unlike a seaweed, plants can't absorb these from the surrounding water; instead, they must be extracted from the soil below by a system of underground roots.

On a warm day, if there's plenty of water, the guard cells around the stomata in the cuticle swell, opening the holes in the leaf. Inside, the leaf is surprisingly empty, with large gaps between the cells that allow gases to diffuse around freely, so carbon dioxide moves into the leaf from the air, replacing the molecules used up by photosynthesis, while oxygen and water tend to leave. The cells in the upper layer of the leaf are crammed with green chloroplasts, all working at full pelt, and they stockpile sugar, some of which is turned into starch grains for longer-term storage and some of which is used in aerobic respiration to recharge ATP batteries. But although the leaf has plenty of sugar, the root cells are condemned to growing in darkness and are in danger of starving to death. Sugar can't be sent to the roots through the xylem, as this only conducts water upwards, not downwards, so the plant needs to invest in a second set of pipes.

Called phloem (pronounced: *flow-em*), the second set of vessels are very different from the water-conducting xylem. Phloem vessels are made from living cells, which are joined together end to end by perforated membranes. The flow of material through the phloem isn't well understood, but we know that sugar can be transported from parts of the plant where it is being made or stored to parts of the plant that need it. In the early spring, the phloem conducts sugar upwards, from storage cells in the roots to the newly developing leaves, while during the summer, the storage organs are replenished, ready to fuel the next year's growth.

The underground storage organs of many plants are an important source of food to a whole suite of animals, including us. Although most human populations around the world rely on grain crops, like wheat and rice, they also eat swollen tubers and roots, like carrots, parsnips, yams and sweet potatoes. Yams are staple foods in many African countries, but unlike grains, yams and potatoes mostly contain starch and lack protein. To get a balanced diet, people need a source of protein to accompany roots and tubers, which means that people can be malnourished even when yams are plentiful.

Plants, like algae, have to make everything they need from inorganic molecules (like CO_2), rather than simply ingesting ready-made building blocks from other cells, as animals do. To make proteins and DNA from the sugary products of photosynthesis, the plant needs other elements – particularly nitrogen and phosphorus, neither of which are easy to find in the world's soils. The hard-working roots explore the spaces between the soil particles, forming a branching network that absorbs mineral ions, often by actively pumping them into the root cells using glucose supplied by the leaves; but mining minerals isn't easy, and plants rely heavily on other organisms to help them out.

Fungi are often considered to be the arch-enemies of plants. Most plant diseases are caused by fungi, and they are one of the few organisms that can digest wood, with some able to eat their way through the hearts of living trees (which is why trees plug up dead xylem). Most of us only encounter the reproductive parts of fungi – the mushrooms and toadstools that we see above ground and sometimes eat – but most fungi are hidden from view, forming fine networks of branching filaments in soils. Fungi might seem unstoppable, but they have a crucial weakness – like animals, they rely on organic carbon obtained from the cells of other organisms and are unable to suck CO_2 from the atmosphere to build their own sugars – so they are on the look-out for partners as well as prey.

In the Rhynie cherts, some of the cells of the early land plants contain strange structures that look like rubber gloves with blown-up fingers. The structures belong to a fungus and it's reasonable to imagine that the fungus is doing the plant harm – perhaps invading the cells to steal sugars, which turns out to be true – but the plant is gaining something essential in return. Natural selection usually demands that organisms are self-interested because it's hard for truly self-sacrificing behaviour to evolve (like Galahad: the doomed saviour of wildebeest). But, interestingly, co-operation between very different organisms is widespread, and plants have struck up associations with all kinds of other living things. These relationships are called *mutualisms*, to emphasize that both parties directly benefit from co-operating with each other, and they occur because all organisms have strengths and weaknesses.

One example of a truly ancient mutualism occurs between plants and a particular group of fungi called *mycorrhizae* (pronounced: *my-co-rye-zee*). Mycorrhizae are exceptionally good at extracting mineral ions, like phosphorus, from soil, but they can't pull carbon out of the air. On the other hand, plants are not very good at extracting phosphorus from soil, but they have fantastic factories, called leaves, that stockpile carbon. If plants and mycorrhizae team up, then each can provide the other with something that they themselves find hard to produce, so both sides win. Indeed, given that this mutualism already exists in the Rhynie cherts, it's entirely possible that the conquest of the land by plants could not have happened without it.

Both fungi and plants exert some control over the relationship. The fungi are normally connected to more than one plant, so they can choose to supply phosphorus to the plant that offers them the best price – paid in carbon. The plants are also normally invaded by multiple different fungi, all offering phosphorus, and so they too can choose which fungus to buy from. This is very similar to many human markets where there are multiple buyers and sellers: the buyers are free to buy from any seller and the sellers are free to sell to any buyer. But how are the prices set?

In a human market, the price is set by supply and demand. If there are lots of buyers and not many sellers, then the price goes up, but if there are plenty of sellers, and not many buyers, the price goes down – and it's no different for plants and fungi. In a soil where phosphorus is scarce, the fungi can demand a higher carbon price, but if phosphorus is abundant, the plants might abandon their fungal partners altogether. The system is stable because the market prevents greedy sellers from charging too much or mean-minded buyers from paying too little: if a fungus demands too much carbon for the phosphorus being offered, then it won't find any buyers; and similarly, a plant can't get away with paying too little for the phosphorus offered by the fungi, as some other plant will be willing to pay more. So, the market helps to prevent cheats from collapsing the system, which seems to have worked for the last 400 million years.

A Free Market

If there are many sellers and few buyers, the sellers won't get a good price.

But if there are few sellers and many buyers, the sellers can overcharge the buyers.

But if there are lots of sellers and lots of buyers, the prices should be fair.

Mycorrhizal fungi and plants take part in a free market. The fungi provide phosphorous (P) and the plants pay for it with sugar.

Each plant is connected to many different fungi and each fungus is connected to many different plants. This keeps 'prices' fair!

Unfortunately for plants, fungi aren't very good at picking up nitrogen, an element that plants desperately need and is normally in short supply. Most of the nitrogen on Earth is in the atmosphere, but fixing atmospheric nitrogen (N_2) into a form that organisms can use requires lots of energy and an environment entirely free of oxygen, so it doesn't sit well within a plant that carries out photosynthesis. But some plants, called legumes, have managed to persuade bacterial master-chemists to do it for them, in return for carbon and an oxygen-free home to live in.

Legumes are a group of plants including peas, beans and lentils that evolved after the end-Cretaceous mass extinction finished off most of the dinosaurs. Humans value and grow legumes because they contain more protein than most plants, which is ideal if you don't have access to meat or fish, or if you are following a vegetarian or vegan diet. Making proteins requires nitrogen, and legumes contain more nitrogen than typical plants because of specialized *nitrogen-fixing bacteria* that live inside their roots.

Nitrogen-fixing bacteria contain unique enzymes to convert nitrogen gas from the air into ammonia (NH_3). Plants can use ammonia and so are keen to recruit these bacteria to help them out. To alert the bacteria to their presence, the roots of legumes secrete a special signalling molecule that acts like an invitation, and the bacteria reply by releasing a signalling molecule of their own.

When the plant receives the signal, it sends out a tiny root and allows a single bacterium to enter. The bacterium multiplies up by gorging itself on the plant's sugars, forming a small round nodule, containing thousands of its descendants. As the nodule develops, the plant manufactures a protein that mops up oxygen, similar to the haemoglobin inside red blood cells, creating conditions under which the bacteria's nitrogen-fixing enzymes can finally get to work. The fall in oxygen triggers the bacteria to switch on genes for nitrogen fixation and start delivering ammonia to the plant, in return for sugars that the plant supplies. But this isn't a free market. A plant usually has lots of different bacteria fixing nitrogen, each housed in its own nodule, so there's lots of sellers, but only one buyer. So how does the plant stop the bacteria from taking the sugar and not fixing nitrogen in return?

To prevent cheating, the plant must police its bacterial partners. To do this,

During the **Carboniferous,** oxygen **levels** peaked at 35% allowing **giant** arthropods to **evolve**

it monitors each nodule and makes sure that each one is providing a reasonable amount of ammonia in return for the sugar that it has supplied. If one strain of bacteria cheats on the plant and doesn't supply ammonia, then the plant destroys that nodule and the bacteria within, while continuing to feed the bacteria in other nodules that are fulfilling their part of the bargain. So, if plants can't rely on the free market to prevent cheating, they resort to intimidation and violence (don't make the mistake of thinking that plants are motivated by peace and love just because they look pretty).

So, plants really are fundamentally different from animals. Plants do not rely on engulfing other cells and rearranging their molecules to build their own bodies; instead they build themselves from scratch. To do so, plants mastered the ability to convert carbon dioxide from the atmosphere into sugar and so gained control of the key commodity of the oxygen-rich aerobically respiring modern biosphere. Then – like any great trading nation – they used their monopoly of this essential resource to drive hard bargains with others. Indeed, plants, despite being rooted to the spot, are the greatest traders in all of life on Earth, and as they continued to evolve, they sought out new partnerships – like the one between legumes and nitrogen-fixing bacteria.

We already know that cyanobacteria changed the world forever when they invented photosynthesis and filled the air with oxygen, so how did the conquest of the land by plants affect the world we live in?

The Rhynie cherts date from the start of the Devonian period. This began around 130 million years after the Cambrian explosion filled the oceans with animal life and lasted roughly 60 million years. The ancient world of the Rhynie cherts would look truly alien to us today, as the land contained only knee-high plants that lacked proper leaves and the only animals were tiny arthropods. But by the end of the Devonian, life on land looked more familiar – at least from a distance.

A treeless world is unimaginable to most of us, but trees only made their first appearance in the middle of the Devonian. Trees don't belong to just one group of plants; instead, the word can be applied to any plant supported by a lot of dead material that holds its leaves high up above the ground. Height gives plants a huge competitive advantage because sunlight always comes from above, so the best way for a plant to shade out competitors is to hold its leaves above theirs and intercept the light – forcing plants to keep up with the neighbours by getting taller and taller. Larger plants also send down deeper roots, so they can access water out of reach to shorter plants, and these combined advantages mean that trees have evolved multiple times and tend to be the dominant plant life form.

During the *Carboniferous* period that followed the Devonian, forests spread throughout the vast supercontinent of Pangaea. Some of the trees that dominated these forests belong to a group called the *lycophytes* (pronounced: *like-oh-fights*), and they were quite different from modern trees in several ways. Despite being large, with trunks up to two metres in diameter, they were mostly supported by bark, rather than wood. They also had strange ways of distributing sugar around their bodies, with phloem tissue that doesn't look like it's quite up to the job. But, the strangest thing about Carboniferous forests is that many of these unusual trees never rotted away. Instead, they lie buried and compressed into enormous coal deposits around the world – giving the Carboniferous (which means coal-bearing) its name.

Coal is simply fossilized plant material. Most plants don't become coal, and it's still something of a mystery why nearly all of the world's coal formed during the Carboniferous. Indeed, if all plants *did* turn into coal, then we would be in trouble. When plants photosynthesize, they lock up carbon dioxide in their bodies, and if this is not eventually returned to the atmosphere, then ecosystems would grind to a halt.

Carbon dioxide in the atmosphere is replenished by aerobic respiration. Most of the sugars made by a plant are respired away during its lifetime, and this respiration can take place inside the plant, or within another organism: the plant supplies sugar to fuel both its own busy roots and the mycorrhizal fungi that mine the soil for phosphorus, while some of the plant's leaves are eaten by energy-hungry animals. Woody parts, like tree trunks, take longer to break down – but even the strongest trees are eventually reduced to gas by hard-working fungi. So, carbon dioxide leads a busy life: a single molecule released by respiration typically only remains in the atmosphere for around eight years, before it is turned back into sugar again by a plant.

So much for short-term exchange, but plants are involved in longer cycles too. A very small percentage of the sugar produced by plants is not respired at all, but ends up buried, deep within the Earth. This is particularly likely to happen when plants grow in waterlogged soils where oxygen can't penetrate, preventing aerobic respiration. Buried organic carbon might not seem like exciting stuff, but it's probably the source of much of the oxygen in the Earth's atmosphere. During photosynthesis, a plant releases one molecule of oxygen to the atmosphere for every molecule of carbon dioxide that it turns into sugar. If that sugar molecule is later respired, then an oxygen molecule from the atmosphere is consumed, and we're back where we started; but if that sugar molecule is forever buried, then the oxygen molecule survives in the atmosphere, allowing oxygen levels to increase.

When plants colonized the land, the concentration of oxygen in the atmosphere jumped up and the concentration of carbon dioxide in the atmosphere nosedived. This dramatic shift was due to the spread of the first global forests, as they drew carbon dioxide out of the atmosphere on a massive scale and incorporated it into their bodies, burying it forever within the corpses of the strange trees of the Carboniferous – or at least until humans arrived. Coal is called a *fossil fuel* because it is made from the bodies of plants that lived hundreds of millions of years ago. Oil (petroleum) and natural gas are other fossil fuels, made from the bodies of small marine organisms, although we are less certain about their origin. Burning is chemically similar to aerobic respiration – so burning coal today returns carbon dioxide to the atmosphere that was first removed by the trees of the Carboniferous hundreds of millions of years ago.

Since the Industrial Revolution, modern humans have relied on fossil fuels to transform their societies, but at the cost of raising global temperatures. Carbon dioxide is a so-called *greenhouse gas* that traps heat in the atmosphere and prevents it from escaping into space. This means that average global temperatures are closely linked to the amount of carbon dioxide in the atmosphere, and by burning fossil fuels and destroying forests, humans have increased the levels astonishingly quickly. Before the Industrial Revolution, there were around 280 molecules of carbon dioxide in every million molecules of air, but by January 2023, this had leaped up to around 419 – and the average global temperature had risen by around 0.8°C as a result.

This rise in global temperatures caused by humans is called *global heating*, and it's

The Impact of Buried Organic Carbon

During most of Earth's history, there was a balance between photosynthesis and respiration, so the concentration of CO_2 and O_2 in the atmosphere remained constant.

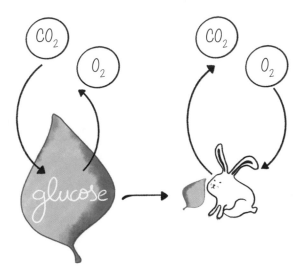

But during some periods of Earth's history, large amounts of plant material were forever buried, leading to an increase in the amount of oxygen in the atmosphere and a decrease in CO_2

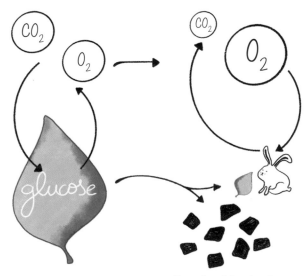

Buried and fossilised

one reason why we need to find alternatives to fossil fuels and capture the carbon dioxide emitted while we continue to burn them.

The lycophyte trees of the Carboniferous were eventually edged out by a group called the **seed plants**. These include **conifers**, like pine trees, which produce seeds in cones and are common in parts of Northern Europe. But although conifers dominate large areas of forest in the higher latitudes of the northern hemisphere, there aren't really very many species, and they only make up a tiny fraction of the world's plants. Instead, most plants on today's Earth belong to a second group of seed plants called the *flowering plants* who made their first appearance during the time of the dinosaurs. The speed with which they diversified and rose to dominance is truly breathtaking, and they now make up around 90% of all plants. Familiar to anyone living in today's world, flowering plants bring welcome bursts of colour and they support vast numbers of pollinating insects, while their seeds and fruits have become essential food for many animals, including us.

The first flowering plants appear in the fossil record during the Cretaceous – the third and final period of the Mesozoic – when dinosaurs were still firmly in charge. But once they appeared, flowering plants rapidly displaced their competitors. Indeed, they took over the world so rapidly that Darwin called their explosive rise an abominable mystery. So, what's so abominable about the rise of the flowering plants?

The hallmark of flowering plants is a complex structure, called a flower, that differs dramatically from the reproductive structures of other seed plants. Peer into a flower and you can usually see both male and female parts. The **stamens** produce pollen that contain the male gametes, and they are usually arranged around a female structure called a *carpel* that contains the eggs. For plants to reproduce sexually, a pollen grain must be transferred to the top of the carpel, where it germinates and forms a tube. The tube grows down to deliver the non-swimming sperm cells to the eggs that lie deep inside. Once fertilized, these eggs become seeds – each containing a tiny plant embryo – and the surrounding carpel swells to form the fruit, which will attract animals to come and eat it and disperse the seeds.

The Structure of a Flower

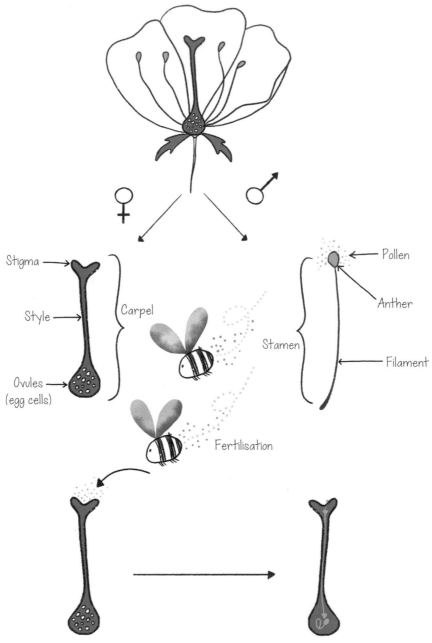

Stigma

Style

Ovules
(egg cells)

Carpel

Pollen

Anther

Stamen

Filament

Fertilisation

Pollen transferred to Stigma

Pollen tube grows down the style
and delivers male gamete

Evolution is a gradual process, so a change of this magnitude – the appearance of a brand-new complex structure – can't just happen overnight. So, how do we explain the apparently explosive appearance and success of the flowering plants? While no convincing fossil of a flowering plant is known before the Cretaceous, comparisons of flowering plant genomes with those of their closest relatives seem to indicate that they do have a much earlier origin. The trouble with the fossil record is that it's inevitably incomplete. Perhaps the earliest flowering plants spent a long time living in an unusual environment where plants don't fossilize well, so it's like looking for a fossilized needle in a haystack of unrelenting rock. But, there's always a chance that some lucky fossil hunter of the future might find the equivalent of the Rhynie cherts for early flowering plants and put the abominable mystery to rest.

While their origins may still be unclear, we can perhaps better understand how flowering plants became so successful. For any group to dominate the Earth, they need to have extraordinary flexibility. We have already seen that the arthropods – the most successful animal phylum – have evolved an incredible array of mouthparts, allowing them to eat practically anything, and the flowering plants also display mind-boggling diversity. Enormous tropical trees, bamboos and grasses, orchids, palms, tiny duckweeds that float on the surface of ponds, succulent cacti that can survive for years without rainfall, and even carnivorous green beasts that feed on unlucky insects – all of these extraordinary green beings belong to the flowering plants. This enormous diversity means that they have occupied just about every habitat on Earth – indeed they truly are those habitats – after all, what is a rainforest without its trees?

The incredible rise of the flowering plants went hand in hand with an explosive increase in the number and diversity of insects. Many plants, like the snails in the garden, are hermaphrodite, so they can pollinate themselves, but sex with others is just as valuable to plants as it is to animals. Transferring pollen from one plant to another is clearly a tough ask for any organism that's rooted to the spot, but plants certainly didn't let that stand in their way. Once again, they evolved a way to get what they wanted through bribery, using their control of the sugar commodities market to manipulate the greedy arthropods who lived alongside them.

Around 90% of flowering plants rely on insects to transfer pollen from one flower to another. The ancestors of flowering plants relied on the wind to carry their pollen, but the wind doesn't always provide a reliable delivery service and it can't be controlled. Plants that depend on wind pollination have to produce extraordinary amounts of pollen – which is wasteful and causes hay fever in unlucky humans – and trust to luck that at least some of it arrives at the right destination. But if plants can persuade insects to visit and transfer pollen, then they can potentially develop their own private courier service for targeted delivery to their friends and neighbours. But this delivery service doesn't come for free.

To persuade insects to visit their flowers, plants secrete a sugary solution called *nectar*. Some plants make it easy for insects of any species to come and help themselves to the nectar in the hope that they will cover themselves in pollen while drinking and so transfer the male gametes to another flower. But a few plants have gone to extraordinary lengths to ensure that their flowers can only be visited by a select few. Indeed, some plants rely on a single species of pollinator. The advantage of having a specialist pollinator is that these insects are much more likely to be carrying the right kind of pollen as they travel around seeking food.

Plants have employed a few different tricks to keep out undesirable insects. One method is to make a flower where the petals are fused together into a very long tube, called a spur, with the nectar buried at the bottom. Only an insect with a very long tongue can now reach the nectar, and these are often moths or butterflies, whose hairy bodies are perfect for picking up and transferring pollen. Darwin was once sent a package of orchid flowers from Madagascar, whose spurs were an astonishing 30 centimetres long, leading him to speculate that one day an insect would be found suitably equipped with a 30-centimetre tongue to reach the nectar. Sure enough, Alfred Russel Wallace, co-discoverer with Darwin of natural selection, described the African sphinx moth in 1867, whose 30-centimetre-long coiled proboscis can be used to reach the orchid's nectar and presumably act as pollinator too – a beautiful example of co-evolution in action.

A few flowering plants have given up on insects and gone back to wind pollination. One of these groups is the **grasses**, an extremely successful group of plants that only evolved during the Cenozoic, once the dinosaurs had disappeared.

The Astonishing Tongue of the African Sphinx Moth

Long spur with
nectar at bottom

Grasses evolved during a period when the continents began to dry out and they now dominate areas where there isn't enough rainfall to support forest cover. In these vast open habitats, the wind can probably do a decent job of transferring pollen, and perhaps the costs of producing more pollen are offset by the savings gained from no longer having to produce nectar to attract greedy insects. The evolution of grasslands in turn supported the evolution of fast-running mammals, like horses, that came to rely on the grasses for food. Grasses can sustain such relentless grazing because their actively dividing cells are buried at the base of the grass-stems, rather than at the tips (as in most plants), allowing them to grow back very quickly when their tips are grazed (or mown) off.

Whether transported by insects or by the wind, once pollen has been successfully transferred to the carpel, fertilization takes place and a tiny embryo begins to take shape inside the developing seed. Unlike an animal embryo, the developing seed will never be able to walk away from its parent, so once again, plants have had to press-gang animals or the wind to help them out. When animals are involved, the carpel swells to form a fruit, packed, inevitably, with the sugar that plants are so free in handing out. Fruits attract animals to take them away and eat them but the seeds inside are normally coated with a tough indigestible covering that protects them as they pass through the animal's gut, allowing them to be deposited unharmed, and encased in some ready-made fertilizer, some distance from the parent plant. The effectiveness of this protective layer explains why tomato plants are often found growing around human sewage farms. However, some seeds don't want to be swallowed by animals and use chemical protection to make them unpalatable – apple seeds are filled with cyanide to dissuade animals from eating them, and so you are advised not to swallow too many of them.

The flowering plants and their sugar factories control the modern biosphere and are essential to all animals. It's therefore strange that plants, with their extraordinary talents, seem to inspire such indifference in many humans. It's true that plants don't move nearly as quickly as animals, and they don't have faces – so we don't know what they're thinking – but their impacts on the Earth couldn't be more dramatic. The greening of the continents changed the atmosphere forever, providing more oxygen for energy-demanding animals, and creating opportunities and habitats for them to exploit. And ultimately, we are just another animal living on this Earth. We too have to fit in to Earth's ecosystem, and we might do well to remember that we are entirely dependent on the bounty of the green things around us.

Chapter 10

ECOLOGY

A brave new world

As the dust settled on the mass extinction that ended the reign of the dinosaurs, modern ecosystems began to emerge. Starting on Boxing Day (26 December) in our one-year timeline, the period between the asteroid impact and the present day is called the Cenozoic, or the era of recent life. Although brief in comparison with other eras of life on Earth, the Cenozoic still spans 65 million years, and during this time two familiar groups of vertebrates, the mammals and the birds, diversified and spread across the globe.

Modern ecosystems are dominated by algae in the sea and by flowering plants on land. Together, plants and algae are often referred to as *primary producers* because they turn raw materials, like carbon dioxide, into sugar, and so provide the key resource on which all other life depends. The ecosystems supported by primary producers can be immensely complex, involving thousands of species that live together and interact by eating each other, competing for food or helping each other out. Our understanding of how ecosystems work is inevitably limited, but the natural world poses many fascinating questions, such as why some ecosystems, like tropical rainforests or coral reefs, contain far more species than others.

The study of plants and animals in their natural environments is called *ecology*. Ecologists track individuals of all kinds of species to find out what they eat, where they go and, crucially, why some survive and thrive while others meet untimely deaths. This information is often combined with mathematical models to help us predict how a population might respond to a sudden change, such as an extreme weather event or the introduction of another species. In fisheries, we might use models to identify whether the number of fish being removed from the oceans is likely to lead to population collapse, while in forestry, we might want to predict how much damage a wood-boring insect will inflict.

Ecological studies have revealed that some species are common while others are rare. Many animals live densely packed together, leading to regular fights over food or nest sites, while others are kept sparse by predators or disease. As well as differences in the average numbers, the populations of some species are stable, while others boom and bust and a few go smoothly up and down with surprising regularity. These differences matter and lead species down different evolutionary paths. For example, a species that regularly gets knocked down to a few individuals is usually able to bounce back rapidly and so is more likely to recover from catastrophic events.

One of the greatest challenges faced by any species on today's Earth is how to cope with an awesome, yet terrifying, newcomer. Modern humans, belonging to the species *Homo sapiens*, have spread around the world and can be found on every continent and in every environment, from tropical rainforests to the Arctic tundra, and they have modified ecosystems more dramatically than any other organism in the history of life on Earth. Indeed, so profoundly have their activities reshaped the world, that many scientists compare their potential impact to that of the asteroid that left the Mesozoic in ruins.

Whether or not this extraordinary life form can rein in its activities and so prevent another mass extinction is surely a burning question for every reader of this book. But before we get there, let's take a look at the ecology of some of the species that surround us to see how their activities have built the ecosystems of the modern world.

Spanning more than three metres, the stiff wings of a *wandering albatross* allow it to glide effortlessly above the mountainous waves in the treacherous Southern Ocean. Albatrosses angle their wings to take advantage of shear forces created by the slowing of the wind near the surface of the water, and so swoop around the oceans at incredible speed and with incomparable efficiency. But there is no room for unforced errors: if an albatross were to misjudge the distance and let its wings awkwardly graze the wave-crests, it would meet its doom in the cold grey water.

Fortunately, this bird is an old master and won't be so easily caught out. At 30 years old, she has circumnavigated the Antarctic continent many times, searching for the nutritious squid that keep her fed, and if she spots a potential meal, she will swoop down and gorge herself – sometimes to the point where she will struggle to take off again. In these terrifying waters, every opportunity must be fully exploited.

Moving out of sight, she makes for Kerguelen, a small island where she has built her nest alongside those of hundreds of other albatrosses. Her landing is clumsy and the long thin wings that served her so well at sea have to be carefully folded back on themselves and tucked away. Gazing around the crowded cliff-tops, she finds that she has landed in just the right spot, and her partner throws back his head and greets her with raucous enthusiasm. They have been reunited every second year for the past two decades, and each time they have managed to rear a single fluffy chick to independence.

Once the chick leaves the nest to embark on its own life, its parents take no further interest, but concentrate on rebuilding their resources for their next breeding attempt. In the years between breeding, these two adults have very different strategies: he hangs around close to the island, while she gradually wanders all the way around Antarctica, seeing different sights every day. But they are a united couple and always pair up again when they return to breed.

Wandering albatrosses don't start to breed until they are at least 10 years old. So, after breeding every second year for 20 years, our middle-aged pair have contributed 10 young birds to the population. If all 10 survive to breeding age and other adults have been equally successful, then the population of wandering albatross should be booming. But in reality, it's in slow decline. So, what exactly determines the fate of populations?

The number of individuals in a population is a balance between births and deaths. For a population to be stable, births and deaths must cancel each other out, so the adults have to raise enough offspring during their lifetimes to replace themselves when they die. If the adults do better than this – and the number of births exceeds the number of deaths – then the population will grow, while the reverse is true for a population in decline.

A simple calculation shows how this works. Some animals are famously prolific, and none more so than the familiar rabbit. Unlike the albatrosses, who lavish their attention on a single chick, rabbits give birth to *litters* of between one and 14 babies. To keep the maths simple, let's assume that rabbits have exactly two litters of six babies before dying themselves, so there will be 12 births for every two deaths.

The State of Populations

A population is STABLE if the number of new adults joining the population is roughly the same as the number of adults that die.

Adults either survive or die

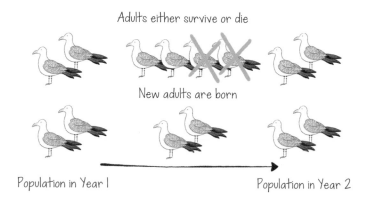

New adults are born

Population in Year I   Population in Year 2

A population will DECLINE if, in a typical year, the number of new adults is smaller than the number of adults that die.

Adults either survive or die

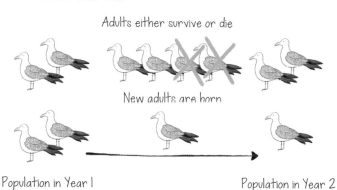

New adults are born

Population in Year I   Population in Year 2

A population will INCREASE if, in a typical year, the number of new adults is greater than the number of adults that die.

Adults either survive or die

New adults are born

Population in Year I   Population in Year 2

Incredibly, after just nine years of playing this particular version of rabbit Happy Families, an initial pair of nose-twitching bunnies would be the proud ancestors of more than 10 million descendants, and it's this inbuilt capacity for growth that gives all species the potential to cover the Earth with their progeny.

These numbers might seem unbelievable, but when British settlers arrived in Australia, they inadvertently tested whether rabbits really were worthy of their reputation. In 1859, the same year that Darwin's *On the Origin of Species* was published, 13 European rabbits were released onto a private estate near Melbourne because its owner had enjoyed shooting them back home. In 1866, hunters shot 14,000 on the same piece of land, and by the 1940s there were an estimated 60 million rabbits across the Australian continent. Various efforts were made to stop their spread, including the construction of a famous fence, around 3,000 kilometres long, which completely failed to contain them. But, this thoughtless experiment confirms that all species really do have the potential to increase in an apparently unlimited way.

Eventually, of course, this unchecked growth is reined in. Rabbits can only rear 12 young per year if they have enough food to eat, so all populations eventually bump up against the limits of their environment. Food is one obvious factor that limits the size of populations, and as food becomes limiting, the death rate will rise or the birth rate will fall, preventing the population from growing any further. Food isn't the only thing that limits populations – for a bird like the wandering albatross that breeds on small remote islands, it might be the lack of nest sites that ultimately curbs growth.

Once species reach their environmental limits, the population stops growing, but individuals are still dying and being born. The result is intense competition, as they battle it out for the few remaining blades of grass or the last vacant nest site. Under these conditions, natural selection will favour features or behaviours that help individuals to win any aggressive encounters, but to secure their fitness, their offspring must also be winners.

In a highly competitive world, animals can ensure that their offspring will succeed by giving them the best possible start in life. Well-fed, healthy offspring are much more likely to win contests over food and nest sites, and so produce the all-important grandchildren that natural selection has made their parents devoutly wish for. But the downside is that each offspring requires massive parental investment, and this limits the number of offspring that their parents can raise. Indeed, the need to produce a heavyweight winner is precisely why the wandering albatross parents only rear a single chick every two years and devotedly feed it until it weighs more than they do.

Many of the charismatic species that humans admire, like whales, elephants and the great apes, only produce small numbers of pampered offspring during their lifetimes. They are all adapted to live in environments where competition is intense, and so producing large numbers of tiny uncompetitive offspring would be a poor strategy. But lavishing attention on a small number of offspring leaves these animals uniquely vulnerable. If they suffer some catastrophe, then once reduced to a few individuals it will take them a very long time to recover their numbers. Indeed, many great whale species – which were decimated by whaling fleets during the 18th and 19th centuries – have still not recovered their former population sizes, despite decades of protection.

At the other end of the spectrum, life for some species is all about taking advantage of opportunities as and when they arise. A good example is the humble greenfly, a small insect that sucks the sap of plants and annoys gardeners by doing so. In places like Britain, the winter is a pretty poor time to be a greenfly, as most plants shed their green leaves and hunker down. Still, some greenfly manage to survive, hidden away in sheltered spots waiting for spring to return, when food suddenly burgeons in astonishing quantities as trees sprout new leaves and plants resume their growth.

To take advantage of this sudden bounty, greenfly populations need to rapidly expand. So, unlike the albatross, natural selection has ensured that greenfly produce enormous numbers of tiny offspring in their short lives, and amazingly, young greenfly are sometimes born with babies already developing inside them. This ability to rapidly increase in numbers makes greenfly populations robust to sudden catastrophe. All animals lie somewhere on a scale between greenfly and albatross. For example, rabbits are sensitive to cold weather, and cold winters can knock rabbit numbers down hard. In the spring following a hard winter, the survivors have a chance to outbreed their competitors and get more of their genes into the next generation, so rabbits start breeding when only six months old and produce multiple large litters.

The Greenfly-Albatross Continuum

In populations that are stable on average, some barely wobble, while others fluctuate wildly

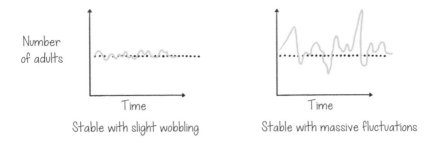

Stable with slight wobbling   Stable with massive fluctuations

Crucial differences in biology underpin these dynamics

At one extreme is the wandering albatross

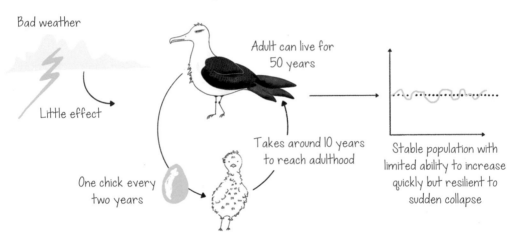

Bad weather

Little effect

Adult can live for 50 years

Takes around 10 years to reach adulthood

One chick every two years

Stable population with limited ability to increase quickly but resilient to sudden collapse

At the other extreme is the greenfly

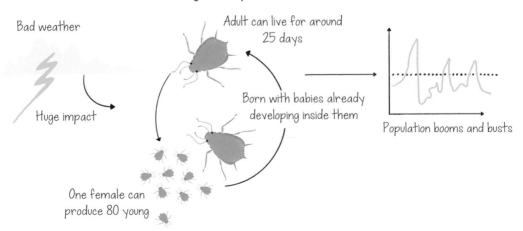

Bad weather

Huge impact

Adult can live for around 25 days

Born with babies already developing inside them

One female can produce 80 young

Population booms and busts

Features of species that are linked to population growth, like the typical number and size of offspring, are called *life-history traits*. Depending on the environment, these traits are under pressure from natural selection to evolve in a particular direction. For the albatross, in its highly competitive environment, this means long-lived adults and small numbers of offspring that are pandered to by their doting parents, while for the rabbit, in its more precarious world, it means living fast and dying relatively young. But it's not just the weather that rabbits have to worry about.

In the forests of the Canadian high Arctic, a relative of the European rabbit, the snowshoe hare, lopes warily through the trees. With enormous back feet that prevent it from sinking into deep snow, the hare changes colour from snowy white in winter to brown in summer. But it's not just the hare's fur that changes rapidly with the seasons; availability of food varies greatly too. During the summer, food is plentiful, with rich new grass and other young plants to eat, while in winter, hares are forced to nibble on dry bark and twigs.

Stealthily pursuing the hare through the very same forests is its sworn enemy, the lynx. Lynx are members of the cat family, about twice the size of a domestic cat, with black ear-tufts and short stubby tails. Snowshoe hares are the favourite prey of the lynx, and when hares are abundant, a lynx will eat around two hares every three days. But the lynx is not the top predator in these forests. Lynx carry a thick coat of long dense fur, which is highly prized by humans, making lynx – and indeed hares – targets for fur trappers.

During the nineteenth century, fur trappers were active across Canada, and they kept good records of the numbers of animals they caught. When plotted on a graph, it became clear that the number of snowshoe hares cycles up and down, peaking every eight to 11 years with as many as one hundred hares in an area the size of a football pitch, before crashing down to as few as ten. Intriguingly, when lynx numbers are plotted on the same graph, they too show a similar pattern, but the peaks and troughs in their numbers seem to lag behind the peaks and troughs in the numbers of hares, leading ecologists to spend a very long time trying to find out why.

Population Cycles: the Snowshoe Hare and the Lynx

Both populations cycle up and down, but lynx numbers lag behind hare numbers

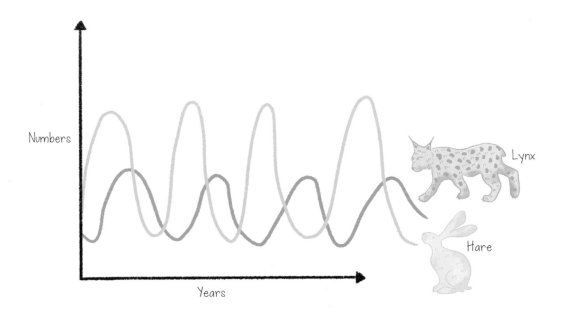

We now think that lynx are partly responsible for the dramatic swings in the numbers of hares, although plants also play their part. As hare numbers peak, nutritious plant food is hard to find and hares begin to starve. But although the hare population may have stopped growing, there are still plenty of hares around, so the number of lynx continues to rise. The starving hares now find themselves surrounded by hungry predators, and the combination of too little food and too many predators causes hare numbers to crash. With too few hares to eat, lynx numbers then inevitably plummet, and the forests echo with the sound of silence. But, with hares reduced to low density, the plants gradually recover, and soon the hare population booms again, dragging the lynx population after it.

The tendency for predator populations to lag behind their prey isn't restricted to hares and lynx. Like the hare, the humble greenfly also has a dedicated enemy. Many gardeners may have noticed that greenfly are a particular problem early in the year, while later in the summer, they seem to have mostly disappeared. Their disappearance is largely due to that voracious greenfly predator, the ladybird, a brightly coloured red or orange beetle with black spots that hibernates through the winter. Once greenfly populations start to rise in spring, ladybirds emerge from their winter sleep and lay their eggs around the rapidly-expanding greenfly colonies. Ladybird larvae are even more voracious feeders than the adults, and consume greenfly in enormous numbers, but just as the lynx population lagged behind that of the hare, so ladybird numbers lag behind their greenfly prey. It usually takes until early summer for ladybirds to get greenfly under control, while during late summer, with greenfly numbers heavily depleted, swarms of hungry ladybirds can sometimes be found wandering around, looking desperately for something to eat.

If predators are so voracious, it's reasonable to ask whether they could actually drive their prey to extinction. When the predator is a *specialist*, and eats only one kind of prey, this is unlikely because the number of predators will quickly decline as its prey becomes scarce, giving the prey a chance to recover. But many carnivores are *generalists*, and will have a go at just about anything. This is problematic for their favoured prey, because even as prey numbers dwindle, predator numbers could remain high if they are supported by large numbers of other prey species that share the same habitat.

We have already seen that various mammals have been introduced to islands by unwitting humans. These are mainly generalist predators, like domestic cats, domestic dogs, rats, foxes and pigs, which eat a wide range of prey, starting with those, like flightless birds, that are easy to catch. People have taken these predators all over the world, including to the continent of Australia, and they have already played a major role in the extinction of more than 150 species (including several flightless birds in New Zealand), not to mention the continuing threat they pose to hundreds of other mammals and reptiles. But predators shouldn't be seen simply as bad guys.

In most places, species within the same ecosystem have a long history of co-evolution, and predators have an essential role to play. An *ecosystem* is defined as all the organisms that live together in one place, including the physical environment, like the

Today,
there are
150,000
sea otters,
half the
pre-hunting
number

soil or water they inhabit. Animals and plants within an ecosystem are connected via *food webs*, which specify who eats whom, and effects in one part of the web can trickle through and have surprising impacts elsewhere, including on the physical environment.

The profound impact of top predators was revealed when fur trappers turned their attention to another North American carnivore, the sea otter. Sea otters are irresistibly cute and they live in coastal waters, feeding on sea urchins, various molluscs and occasionally fish. Because they largely eat invertebrates protected by hard shells, sea otters keep a special rock tucked into a loose pouch of skin on their chest, together with any food they have recently picked up from the sea floor. On returning to the surface, the otter floats on its back, puts the rock on its chest and bashes the unfortunate seafood against the rock to smash it up before wolfing it down.

The hunting of sea otters for their fur began in Russia during the eighteenth century, but soon spread down the west coast of North America, where otters were abundant. The hunting was so intense that otter numbers were reduced to somewhere between one and two thousand, and its extinction was widely expected. A hunting ban saved the species, which has now rebounded, covering around two-thirds of its former range. But during the dark days of exploitation, when its numbers were severely reduced, we learned how predators can be essential for the health of entire ecosystems.

Sea otters inhabit kelp forests. Kelp is a type of brown seaweed – one of those strange organisms that captured a photosynthetic alga and put it to work. Kelp forests are enormously productive and play host to an impressive number and diversity of other organisms: sheltering molluscs, urchins, fish and otters alike. But some of the organisms they host pose a threat to their very existence.

Sea urchins are voracious grazers. Encased in a hard shell, and often bristling with sharp spines, they crawl along the bottom and eat away at the kelp's holdfast, causing the whole thing to float away. If their population booms, their unstoppable appetites can create urchin barrens: areas of seabed scraped clean of all algae, and with it the habitat that would have supported thousands of others. Fortunately, their populations are normally kept in check by predators, and one of their main predators is the sea otter.

By 1911, sea otter populations had been decimated by hunting, and the number of urchins rocketed. Their intensive grazing devastated kelp forests up and down the west coast of North America, leaving the inhabitants with nowhere to go. Thankfully, once otter populations were restored, recovery in many places slowly followed, and otters were dubbed a *keystone species*, defined as species that have far-reaching impacts on the ecosystems they inhabit.

So, far from being the bad guys, predators often influence ecosystems in positive ways. By keeping herbivore populations in check, healthy predator populations prevent plants or algae from being destroyed by overgrazing, and there are ongoing attempts to restore predators to several degraded ecosystems in the hope that they can create a better balance. But for many species, predators probably aren't their main concern. Instead, they need to stay focused on the competition.

Evolutionary biology is rooted in Darwin's theory of natural selection, but ecology also has a central guiding principle. Often attributed to a Russian scientist named Gause, it states that species cannot coexist if they are too similar; instead, one will drive the other to extinction. Roughly translated, this means that we expect species sharing the same habitat to be different in ecologically meaningful ways. So, how does this work in practice?

Barnacles are crustaceans, closely related to shrimps, crabs and lobsters. As adults, they cement themselves to rocks and secrete a small pyramid-shaped shell to protect their vulnerable bodies. When covered by the sea, a hole in the top of the pyramid opens, allowing eight pairs of feathery limbs to emerge, which open and close to trap particles of food which the barnacle then eats. This simple but effective lifestyle has allowed barnacles to become fantastically abundant, and on some rocky shores they appear to plaster every available surface. But – in apparent violation of Gause's principle – more than one species can often be found in the same place.

While it's common to find at least two different barnacle species on a given rocky shore, they aren't all mixed up together. Around the coasts of Britain and North America, one species dominates the upper shore, while a second is restricted to the lower shore. On the upper shore, barnacles are left high and dry for most of the day, while on the lower shore, they are covered by water for longer periods of time, so they can feed more often and grow faster. The larvae of both species seem to settle all over the place, but they clearly don't survive equally well. On the lower shore, the lower-shore specialist takes advantage of the extra feeding opportunities to undercut the upper-shore specialist, by overgrowing it and displacing the developing larvae. But, on the upper shore, the more aggressive lower-shore specialist simply can't survive the long periods of drought, leaving the upper-shore specialist to dominate.

Meaningful ecological differences, like those seen in barnacles, are common in many apparently similar species. We say that each species has its own *niche* (pronounced: *neesh*), defined by a whole range of ecological factors, like food preferences, tolerance of drought and ability to escape from predators. But whether or not two species with different niches can coexist also depends on the environment they inhabit. The rocky shore can host two species of barnacle because the twice-daily tides create an environment on the upper shore that is quite different from that on the lower shore, so each species can find its niche. And the more complex an ecosystem becomes, the more species it can potentially accommodate.

Some places on Earth positively teem with species. These hotspots of diversity are usually found in the tropics. Somewhere between 40 and 50 thousand different tree species occur in tropical rainforests across the world, and if we zoom into just one of those forests, we might find several hundred different tree species in an area the size of a football pitch. In contrast, European forests, while still undoubtedly beautiful, support 124 species of tree in total.

The astonishing diversity of trees in tropical forests supports other life in turn. Whether they want to or not, trees provide homes and food for other organisms. Specialized plants called *epiphytes* attach themselves to their bark; insects bore into their wood and eat their leaves and seeds; and mammals roam through the undergrowth snacking on their fallen fruit. While some tree-dwelling animals are generalists, most of the insects are specialists and can only eat the wood, leaves or fruits of a single type of tree, so a forest with more tree species provides opportunities for more species of animal.

Zonation on the Rocky Shore

The difference between high and low tide on some beaches can be dramatic

Tidal zone

Some animals, such as barnacles thrive in the tidal zone. They emerge to feed when covered by water

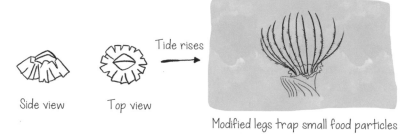

Side view   Top view   Tide rises

Modified legs trap small food particles

Different species can tolerate being dried out for different periods of time and this leads to zonation

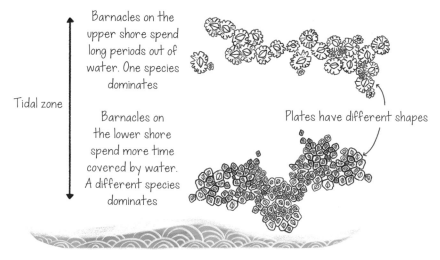

Tidal zone

Barnacles on the upper shore spend long periods out of water. One species dominates

Barnacles on the lower shore spend more time covered by water. A different species dominates

Plates have different shapes

Zonation demonstrates an important ecological princple: no species can be good at everything

Confining yourself to eating the leaves of only a single type of tree seems like a poor strategy for any animal, especially as plants don't make it easy for insects to eat them. Plants stuff their leaves with an enormous variety of chemicals to deter herbivores from feeding, and insects have had to evolve enzymes to detoxify their dinner. Each species of plant produces a different chemical cocktail, and an insect has to have the right number and type of enzymes if it doesn't want to be poisoned every time it tucks in.

Returning to the humble greenfly, we can see how this specialization works. Greenfly are more correctly known as aphids, and they aren't all green. Over five hundred different species of *aphid* can be found in Britain, and their names reveal their dietary preferences: the speckled larch aphid feeds only on larch trees; the willowherb aphid feeds almost entirely on the broad-leaved willowherb (a rather weedy-looking plant); and the ivy aphid feeds only on ivy (a clinging evergreen that covers walls and trees all over Britain). Plant chemistry clearly plays a role in this specialization because when aphids feed on more than one plant species, they are nearly always closely related, so the cabbage aphid can actually feed on many different plant species, but only those within the cabbage family.

The specialization displayed by aphids is an example of Gause's principle in action. For aphid species to coexist, they can't be exactly the same, so there is only one species of ivy aphid, not three. But this principle doesn't set very hard limits on diversity, as there are so many opportunities out there. If a mutation arises that allows an aphid to eat a new plant that doesn't yet have a specialist aphid, then it will probably be successful – potentially setting the mutant aphids on a path to becoming a new species. Similarly, if a mutation arises in a tree species, changing its chemistry so that an aphid can no longer attack it, then this could eventually give rise to a new species of tree.

Diversity in one part of a food web encourages diversity elsewhere. Each specialist aphid is attacked by its own *parasitoid* – in this case members of a wasp family that lay their eggs inside a living host. The larvae develop inside the unfortunate host and once they have consumed its innards, they pupate within its mummified remains, before emerging as a winged adult.

Parasitoid wasps are one of the most diverse groups of insects in the world. Britain alone boasts several thousand species, although its human population is largely unaware of their existence. At just a few millimetres long, most lay eggs on just one species of insect host, which might be an aphid, a caterpillar or indeed another wasp – so they are generally to be welcomed, at least by gardeners and farmers. Their specialization is driven by the intimate nature of their relationship with their victim. First, they need to locate it, which requires the ability to detect very specific chemicals released either by the host or the plant it feeds on. Second, they develop inside another organism, so they need specific adaptations to navigate and process the peculiarities of the host's body. The upshot is that most parasitoids only target one host, giving them plenty of opportunities to diversify into different species.

Specialization causes diversity to build up over long periods of time. Modern ecosystems have evolved over millions of years, but some have had a more disturbed history than others. Starting around 2.6 million years ago, the *Pleistocene* period, or Ice Age, ushered in cycles of glaciation that plunged the Earth into deep cold every 100,000 years or so. Glaciers pushed down from the poles, devastating diversity in northern latitudes and leaving today's European forests impoverished. Even the tropical rainforests weren't immune to the cold, and they shrank back into smaller patches, from which they spread out again whenever the ice loosened its grip. The effect of glaciations can still be seen today: the African rainforests have much lower diversity than those in South America or Asia because in Africa forests were forced back into far fewer smaller patches as the ice extended south.

Despite various setbacks, today's planet hosts an exuberant variety of life. It's difficult to know exactly how much diversity the Earth supports, but the best estimates lie somewhere between 5 and 10 million species. This richness is a product of four-and-a-half-billion years of evolution, as species have diversified and occupied every possible environment. Life may have begun inauspiciously with single cells in a deep-sea vent, but today, planet Earth is surely one of the wonders of the universe. Only one big question remains: while we continue to make progress unravelling the Earth's past, what does its future hold?

By around 20,000 years ago, modern humans had begun to leave their mark.

How Specialization can Lead to High Diversity

In Britian, the oak is one of the most common trees. Trees can live for several hundred years

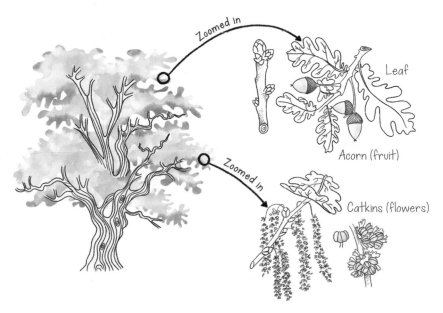

Zoomed in

Leaf

Acorn (fruit)

Zoomed in

Catkins (flowers)

There are 326 species that entirely depend on oak trees

The purple hairstreak butterfly lays eggs on the oak flowers

The jay eats huge numbers of acorns and stashes them for future use

Knopper gall

Oak apple

Oak gall wasps lay eggs in developing acorns which then house and feed their larvae

In tropical forests, there are thousands of different tree species, each with its own community of specialists. The total diversity is staggering

On the walls of decorated caves on every continent, our ancestors painted the world around them. Standing out in glowing colours are enormous beasts – lions, rhinos, bison and giant aurochs (extinct relatives of today's domestic cattle). In firelight, the images must have flickered and danced on the cave walls, and the artists have captured the essence of these fierce and dangerous animals with great effect. In sharp contrast, our ancestors are conspicuous by their absence in the images. Far from putting themselves centre stage, they are pushed to the margins, at least if we assume that the rather badly drawn faceless stick-figures represent the artists or their fellows.

To Stone Age people, perhaps it would have seemed laughable to place themselves at the centre of a world that was filled with magnificent giants. They had little control over their environment and probably saw themselves as an integral part of the natural world, not something special that was set apart. Hunting other animals for food and occasionally falling victim in return – this was how the world worked, for them and every other inhabitant of the planet.

Roughly 12,000 years ago, the period when humans could reasonably be thought of as 'just another species' was ended by the Neolithic revolution. Neolithic people didn't just slot into existing landscapes; instead, they began a process of wholesale transformation, by replacing natural ecosystems with ones that would serve their needs. Of course, we have to admire their determination – to destroy a standing forest with a stone axe is a monumental feat, comparable with anything humans have achieved more recently. But it set people against nature and meant that humans were no longer on the margins of life on our planet.

With the advent of farming, people stopped wandering – moving from place to place following prey or seeking other food sources – and settled in one place where their food was grown or husbanded. Their settlements grew into villages and towns, and their populations expanded. Diverse natural ecosystems supporting thousands of species were converted into fields dedicated to a small number of crops that could feed a permanent human settlement. In Eurasia, the birthplace of agriculture was the Fertile Crescent in the Middle East. Lying between two great rivers in modern Iraq, the Tigris and the Euphrates, Neolithic farmers domesticated wheat and barley from wild grasses, a process that would have taken considerable time and dedication.

Wild grasses produce small seeds that fall to the ground, making them difficult to harvest. To improve yields and make harvesting easier, early farmers were careful about which seeds they saved to sow the following year, selecting only those with desirable characteristics. By choosing seeds from plants that produced lots of large seeds that didn't drop to the ground easily, they gradually improved yields so that more people could be fed. This process, called *artificial selection*, was also applied to animals – domestic cattle, pigs and sheep are all descended from wild animals, but they are more docile and less intelligent than their forebears.

The domestication of plants to satisfy Neolithic appetites wasn't restricted to the Middle East. In the Americas, people bred *maize*, with its glowing corn cobs, from a less attractive wild relative named *teosinte* (pronounced: *tee-oh-sin-tay*), and in Asia, rice was grown in regimented paddy fields. By adding the dung of domestic animals and waging war on weeds and pests, yields were enhanced still further, but agriculture was mainly a subsistence activity that demanded significant investment of human labour.

With the advent of the Industrial Revolution, human workers were replaced with machines, and agricultural production started to seriously scale up. By the mid 1850s, European farmers were fertilizing their fields with bird poo (*guano*) that was scraped by the tonne from seabird colonies around the world. Guano contains nitrogen, phosphorus and potassium – elements that plants need for rapid growth but are normally scarce in soils – and in the early twentieth century, Fritz Haber devised a chemical process to convert nitrogen from the atmosphere into ammonia, and humans began to mine phosphate-containing rocks. Today, these inorganic fertilizers have largely replaced those of biological origin, and yields of wheat have risen from around one tonne per hectare in 1850 to ten tonnes per hectare in 2020.

The consequence of more food, alongside parallel advances in medicine, was a boom in the human population. In the year 1800, the total number of people around the world reached one billion for the first time; by 1930, it had doubled; and by 1974, it had doubled again. In 2011, the human population reached seven billion, and we are on track to hit eight billion by 2023. This seemingly unstoppable growth is partly due to the industrialization of agriculture, including the breeding of high-yielding varieties of crops and the use of inorganic chemicals to fertilize and protect them.

There are microplastics in freshly fallen Antarctic snow

But the price of our unprecedented success is being paid by other inhabitants of our planet. Eight billion people living in technological societies inevitably leave a lasting impression. From plastic flotsam floating across the world's oceans to traces of pesticides contaminating remote food chains, the influence of humans can be seen or felt in every corner of today's Earth. So, to minimize our impact and safeguard future generations, perhaps we need to think more carefully about how humans interact with the rest of the natural world.

It's never a good idea to get too starry-eyed about your ancestors. Sooner or later, a skeleton is bound to come tumbling out of the closet, and in the case of Palaeolithic people, that skeleton turns out to be gigantic. Despite the modesty expressed in their cave art, Stone Age people were formidable hunters, and although scientists still argue about it, they were almost certainly involved in the extinction of the great beasts they so clearly admired.

The size of mammals increased steadily throughout the Cenozoic as the small-bodied survivors of the end-Cretaceous mass extinction diversified. We have already seen how large body size brings many benefits, and by the onset of the Pleistocene, 2.6 million years ago, a fabulous *megafauna* was spread across all continents. Prominent members of the megafauna include sabre-toothed cats, cave bears, dire wolves, lions, mammoths, mastodons and woolly rhinos – and supersizing wasn't restricted to mammals. In New Zealand and Australia, giant flightless birds also roamed, including the famous New Zealand moas, which stood more than three metres high. We know that human hunters tend to target large animals, perhaps because they will feed more people once killed, but large animals are peculiarly susceptible to hunting.

Most populations can tolerate some level of hunting because, if numbers are reduced, the survivors have more food, allowing the population to recover. But the speed with which they recover depends on their life-history traits. Rabbits, with their prodigious capacity to reproduce, are resilient to hunting – indeed, hunters tried to control rabbits in Australia and failed – but an albatross, which only has one chick every two years, is much more vulnerable.

The megafauna were all large-bodied animals with low rates of reproduction – much more similar to the albatross or the blue whale than the rabbit. Although Palaeolithic hunters didn't achieve the population densities of Neolithic farmers, they could probably have killed enough animals to push vulnerable species to the brink. Combine these losses with periods of rapid climate change during the Ice Age, and the stage is set for widespread extinctions. As the glaciers melted for the last time and the Pleistocene drew to a close, most of the remaining megafauna were lost.

So, how can modern humans minimize their impact on the planet they inhabit? First, we should remember that large animals are vulnerable to any level of harvesting or poaching. The wandering albatross that sailed through the opening paragraph of this chapter is an example of an animal currently threatened by human activities. This isn't deliberate – humans no longer hunt albatrosses – but they do hunt fish, by hanging out miles of hooks, baited with squid, to catch predatory fish like tuna. Albatrosses must take advantage of any opportunity to feed, and bits of squid floating at the surface are irresistible – so they swoop down and get caught on the hooks. Inexorably, year on year, populations of albatross around the world are sliding into decline, and if this continues, this magnificent bird is destined to wander only to extinction. Indeed, many of our best-loved and most exciting large animals may unfortunately share its fate, as humans target their populations for food, pets and medicine.

Second, the capacity of our planet is finite. All species eventually bump up against the limits of their environment, and humans are no exception. People in previous centuries couldn't have known that their activities were profoundly affecting a world that appeared to be infinitely large. But today, we know that eight billion people can change the composition of the atmosphere and seriously deplete populations of even abundant species. Learning to live within ecological boundaries is a major challenge, and requires all of us to think hard about how we grow food, generate energy and interact with the rest of the natural world.

Third, humans are astonishingly creative, and we shouldn't just dwell on our destructive potential. There's no reason to imagine that humans can't live alongside nature or that a new mass extinction is inevitable. Indeed, if we could apply the collective energies of the best human minds on our planet to the problems of our time, then surely solutions will not be hard to find. And what about us as individuals?

The challenges that now face our planet can easily feel overwhelming, so it's hard to imagine how we can play a part, or that a very average human being in a very average town has something to offer the world. And if that's how you're feeling, then think about this.

Every person alive today has an unbroken line of ancestors. This ancestry can be traced back through their parents and grandparents to the Neolithic farmers who domesticated plants and animals and set the stage for a technological society. Before that, their Palaeolithic grandmothers must have danced in a decorated cave and celebrated their lowly place in a world filled with unfathomable wonders – little knowing the impact they were having on it. If we go back further still, then their great-grandfathers were apes, and before them, they were insectivorous mammals, busily hiding from dinosaurs.

Perhaps at the start of the Mesozoic, one of these ancestors was a *Lystrosaurus*, who survived the Great Dying, and before that, their intrepid amphibious forebears might have paddled their way around terrifying swamps and battled with armoured fish. At the dawn of the Palaeozoic, their extremely ancient great-grandmothers probably swam in the Cambrian seas and were part of the explosion that fired up the Animal Kingdom. And now we have to reach back to a world before multicellular animals even appeared.

For the three and a half billion years before the Cambrian, the ancestors of everyone alive today was a series of single cells, which become less and less sophisticated the further back in time we go. As we rewind faster and faster, we will witness the unmaking of the eukaryotic cell, as the enslaved bacterium goes back to its free-living state, and the world becomes dominated only by bacteria and archaea. Eventually, we will find ourselves in a deep-sea vent, admiring the ultimate ancestor of all of us – a proto-cell that is learning to pump protons before it moves out to conquer the world.

Unlike religion, science can't offer cosmic significance, but perhaps this thought will bring some comfort. Every one of the trillions of ancestors that we can lay claim to, from that single cell in the deep-sea vent to our current hard-working parents, were winners. They battled through the cut-throat mill of natural selection with the odds stacked heavily against them and emerged, bloody but victorious, to carry on their lineage and give us the chance of a lifetime: to live on the only planet in the known universe that's filled with life; to breathe oxygen-rich air; to admire a bird flashing past or simply watch a humble earthworm escaping from it. What we choose to do with the brief time we have is, of course, entirely up to us. But, if you're looking for inspiration, then what could be better than that?

GLOSSARY
&
INDEX

Activation energy: the energy required to start a chemical reaction. Can be substantially reduced by enzymes.

Adaptation: a feature or behaviour of an organism that makes it able to function well in its environment.

Aerobic respiration: the process by which glucose is converted to carbon dioxide and water with the help of molecular oxygen, liberating a large amount of energy. (See also Anaerobic respiration and Fermentation)

Algae: a broad term that includes many different organisms that live in water and photosynthesize.

Alginates: slimy substances secreted by some seaweeds that reduce water loss.

Altruism: to sacrifice yourself for the benefit of others.

Amino acids: the building blocks of proteins. Twenty different amino acids are used by cells, including arginine, glycine and valine.

Ammonites: a group of extinct squid-like marine animals that produced a coiled shell in which to live.

Amoeba: a large free-living eukaryotic cell that captures and eats other smaller cells.

Amphibian: a type of tetrapod with moist skin that lays eggs in water.

Anaerobic respiration: another word for fermentation (see Fermentation and Aerobic respiration).

Animal Kingdom: a massive group of multicellular creatures that can move, reproduce sexually and harness energy from aerobic respiration.

Annelida: a phylum within the Animal Kingdom that contains worms with segmented bodies, such as earthworms.

Anomalocaris: an extraordinary large arthropod predator from the Cambrian seas.

Antibiotic: a drug that kills bacteria inside the bodies of their hosts without damaging the host.

Antibody: a molecule produced by vertebrate immune systems that locks onto 'non-self' cells.

Antigen: a molecule on the surface of a cell that can be recognized by cells of the immune system and reveals whether the cell belongs to the body in question or is 'non-self' and should be attacked.

Aphid: a small insect that sucks the sap of plants.

Archaea: one of three fundamental types of cell. Often associated with extreme environments, such as hot springs and deep-sea vents. (See also Bacteria and Eukaryotes.)

Arthropods: a phylum within the Animal Kingdom that contains insects, crustaceans (shrimps, crabs and lobsters), spiders, centipedes and many others that we might call creepy-crawlies.

Artificial selection: the process of changing the way a domestic animal or plant looks or behaves by breeding from individuals with desirable features.

Asexual reproduction: a form of reproduction that does not involve sex by producing offspring that are clones of their parents. (See also Sexual reproduction.)

Asgard Archaea: a group of archaea that are believed to be the ancestors of the eukaryotes.

Atoms: the building blocks of all matter in our universe.

ATP (adenosine triphosphate): a small molecule that acts like a rechargeable battery.

Bacteria: one of three fundamental types of cell that evolved very early in the history of life. (See also Archaea and Eukaryotes.)

Banded iron formations: huge deposits of iron oxide that are thought to have formed when oxygen produced by cyanobacteria reacted with iron in the Earth's crust.

Bilateria: a group containing all the animal phyla that have bilateral symmetry.

Bivalve: a type of mollusc with two shells that fit closely together.

Body plan: a term used to describe the overall pattern of an animal's body, including important features like the number of segments, legs and lines of symmetry.

Boring billion: the period between the evolution of the eukaryotic cell 1.8 billion years ago, and the emergence of the first multicellular beings, around 800 million years ago.

Cambrian explosion: a short period of geological time in which modern animal groups appear for the first time in the fossil record.

Capillaries: the smallest blood vessels.

Carbohydrates: a broad term to describe both simple sugars and larger macromolecules, like cellulose, that are made by joining simple sugars together.

Carbon: an element and the basis of all life's molecules.

Carboniferous: a period during the Palaeozoic when towering forests that eventually gave rise to coal covered much of the Earth.

Carpel: the female part of a flower that contains the egg cells.

Cartilage: a flexible, lightweight material used in animal skeletons.

Cell: the fundamental unit of all living things.

Cell differentiation: a process during the development of an animal's body in which cells gradually become specialized to take on particular roles.

Cell division: the process by which a cell splits itself into two daughter cells.

Cell membrane: a flexible sheet made from lipid that surrounds cells and organelles.

Cell nucleus: an organelle found inside eukaryotic cells that contains the genome.

Cellulose: a macromolecule that is used to build plant cell walls.

Cenozoic: the recent era of life on Earth. Runs from 65 million years ago to the present day.

Central dogma: the one-way flow of information from DNA to RNA to proteins.

Cephalopod: a type of mollusc that sprouts arms from its head, such as octopus and squid.

Chaperone proteins: proteins that help other proteins to fold into their proper final shape.

Chert: a glassy rock in which the remains of extinct plants and animals can sometimes be found.

Chlorophyll: molecule at the heart of a photosystem that absorbs the sun's energy and destroys water molecules, liberating hydrogen and oxygen.

Chloroplast: an organelle containing chlorophyll found inside plant cells where photosynthesis takes place.

Chordate: a phylum within the Animal Kingdom that includes all the vertebrates, such as fish, dogs and humans.

Chromosomes: long sections of DNA that carry part of the genome. Species with large genomes require more than one chromosome.

Cilia: short hair-like structures that can cover the outside of cells and beat together in co-ordinated waves.

Circulatory system: a system, including a pump and a network of pipes, that delivers glucose and oxygen to cells within an animal's body.

Cloaca: a general-purpose organ possessed by reptiles (including birds) through which they can excrete various kinds of waste.

Clones: individuals or cells with identical genomes.

Cnidarians: a phylum within the Animal Kingdom that contains jellyfish, sea anemones and corals.

Co-evolution: a tight evolutionary relationship between two species where each evolves in response to changes in the other.

Collagen: a protein with elastic properties that many animal cells secrete and is important in bone, skin and cartilage formation.

Collar cells: cells that line the interior of a sponge.

Colony: a group of similar units, like cells or organisms, that work together but also retain their individuality.

Comb-jellies: a phylum within the Animal Kingdom that contains jelly-animals that swim using thousands of tiny cilia that beat in co-ordinated waves.

Conifer: a type of seed plant that produces seeds inside cones.

Copying machinery: a collection of enzymes that prise DNA strands apart, reveal the genetic code (letters) and then make an identical copy, so that each daughter cell can receive a copy of the genome.

Cost of sex: the reduction in population growth rate that sexual species suffer compared to asexual species, mostly due to the production of males.

Crustaceans: a group of animals that includes crabs, lobsters and shrimps.

Cuticle: a waterproof layer that prevents water loss from plant leaves.

Cyanobacteria: the original inventors of photosynthesis which inhabit the oceans, freshwater and other damp environments.

Cystic fibrosis: a genetic disorder mainly affecting the lungs.

Cytoplasm: the watery fluid inside cells, including any organelles, except the cell nucleus.

Cytoskeleton: a dynamic spiderweb of filaments inside eukaryotic cells that support the cell and allows it to change shape and crawl.

Cytosol: the liquid part of the cytoplasm in which organelles are embedded.

Deep-sea vents: structures on the deep ocean floor where seawater interacts with molten rock within the crust, causing the water to boil. These vents are rich in chemicals and a possible site for the origins of life on Earth.

Devonian: a period during the Palaeozoic in which the tetrapods came onto land.

Diffusion: the passive movement of molecules from places where they are plentiful to places where they are scarce.

Dinosaurs: an extinct group of terrestrial reptiles, including favourites such as *Tyrannosaurus rex* and *Diplodocus*.

DNA (deoxyribonucleic acid): an information molecule that consists of any number of DNA nucleotides (letters) strung together to form a long strand, usually bonded to a second strand to form a double helix.

Domains: scientists currently recognize three domains of life: Bacteria, Archaea and Eukaryota (or Eukarya).

Doubling time: the length of time it takes for a single cell to turn itself into two daughter cells.

E. coli: a common type of bacteria that lives in the guts of humans. Mostly harmless, but a few strains can cause serious illness.

Ecology: the way in which populations of plants and animals are linked together and the study of this.

Ecosystem: all the organisms that live together in one place, including the physical environment, such as the soil or water that they inhabit.

Ectotherm: an animal that takes its temperature from the environment. (See also Endotherm.)

Ediacara: the world before the Cambrian explosion, in which soft-bodied, multicellular creatures that are nothing like modern animals lived.

Egg cell: a female sex cell that is usually large and immobile.

Elastin: a protein that gives lungs and blood vessels their elastic properties.

Electron: a negatively charged particle that orbits the nuclei of atoms.

Electron microscope: a microscope with very high magnification that can image the interior of cells in great detail.

Element: a naturally occurring substance, such as oxygen, carbon or iron, that cannot be broken down into smaller chemical parts. Each element is defined by the number of protons in the nucleus.

Embryo: an early stage of an animal's development during which cells become more specialized and organs form.

Endoplasmic reticulum: the first part of the warehouse inside eukaryotic cells that produces, modifies and transports proteins.

Endotherm: an animal that maintains a constant body temperature through metabolic activity. (See also Ectotherm.)

Entropy: disorder.

Enzyme: a molecular machine that speeds up chemical reactions inside cells.

Eon: the longest slice of geological time (see also Era and Period).

Epiphyte: a plant that grows on another plant but generally does it no harm.

Epithelial cells: a key animal innovation, these cells line the insides and outsides of our bodies, forming flexible, watertight sheets.

Era: the second-longest slice of geological time. (see also Eon and Period.)

Eukaryote: the third fundamental type of cell. An unlikely mash-up between archaea and bacteria. (See also Archaea and Bacteria.)

Evolution: changes in the genetic make-up of a population that are wrought by natural selection.

Exoskeleton: the hardened external covering of an animal, such as a beetle or crab, that protects the soft body within.

Extinct: a species that has died out and can no longer be found anywhere on Earth.

Fatty acids: one of the four fundamental chemical building blocks of cells. Used in cell membranes. (See also Simple sugars, Amino acids and Nucleotides)

Fermentation: a process by which a six-carbon glucose molecule is split into two three-carbon molecules, releasing a small amount of energy. Does not require oxygen (also called anaerobic respiration). (See also Aerobic respiration and Anaerobic respiration.)

Fertilization: the joining together of an egg cell and a sperm cell to form the first cell of a new individual.

Filter-feeding: a mode of feeding used by aquatic animals that involves filtering out particles of food from the water.

Fitness: the value of an organism as perceived by natural selection. Individuals with high fitness are highly likely to survive and produce large numbers of offspring.

Fixing: converting a molecule into a biologically useable form.

Flagellated funnel cells: free-living eukaryotic cells that are thought to be similar to the ultimate ancestor of the Animal Kingdom.

Flagellum/Flagella (bacterial): thin, rigid, corkscrew-like projection/s found on many bacterial cells that allow the bacteria to swim.

Flagellum/Flagella (eukaryotic): a long whip-like extension that protrudes from the cell and flails back and forth, allowing it to swim.

Flowering plants: a type of seed plant that produces elaborate flowers with male and female parts.

Food web: a diagram that shows who eats whom or what in an ecosystem.

Fossil fuel: a fuel made from the remains of long-dead organisms.

Fossils: the remains of plants and animals – or traces of their presence – that have turned to stone.

Gamete: a sex cell that only contains half of the normal genome.

Gene: a sequence of DNA letters that encodes a single protein.

Gene editing: a technique that allows the sequence of DNA letters within a gene to be changed.

Gene expression: the process of switching genes on and off.

Generalist: an organism that eats a wide variety of food.

Genetic disorder: a condition that arises because of inherited problems in a person's genome.

Genome: a blueprint carried by all living things that contains instructions to build proteins.

Genome sequencing: a technique that allows us to read every letter in all or part of the genome.

Gills: organs possessed by many marine animals that allow gas exchange between the water and the blood.

Global heating: the change in the Earth's temperature caused by excess greenhouse gases, which have been mostly produced by burning fossil fuels.

Glucose: a small sugar molecule with six carbon atoms that contains a lot of chemical energy.

Golgi apparatus: the last part of the warehouse inside eukaryotic cells that produces, modifies and transports proteins.

Glycoprotein: a protein with a carbohydrate molecule attached.

Grass: a type of flowering plant that relies on the wind to pollinate its flowers and can withstand heavy grazing.

Great Oxidation Event: occurred around 2.4 billion years ago, when the concentration of oxygen in the atmosphere rose suddenly from practically zero to a measurable level.

Greenhouse gas: a gas, such as carbon dioxide, that traps heat in the atmosphere.

Guano: bird poo often used as a fertilizer, and that accumulates in seabird colonies.

Guard cells: cells that can change shape to open and close the stomata in a leaf.

Hadean: the very earliest eon in the Earth's history.

Haemoglobin: a very large protein that carries oxygen in the blood of many animals.

Heart: a muscular pump that moves blood around an animal's body.

Hermaphrodite: a sexually reproducing individual that can produce both male and female sex cells in the same body.

Holdfast: a structure that anchors a seaweed to the seabed.

Hormones: chemicals that travel through the body and affect the behaviour of other cells.

Housekeeping genes: the genes that run the day-to-day activities in a cell and are always switched on.

Hydra: a simple freshwater Cnidarian with a ring of tentacles.

Hydrogen sulphide: a smelly gas formed when hydrogen reacts with sulphur. Emitted around areas of volcanic activity.

Ichthyosaur: a large extinct marine reptile that resembles a modern dolphin.

Immune system: a specialized system in the bodies of some animals that can recognize invading cells and kill them.

Industrial Revolution: dating from 1760 to 1820, a time when coal-fuelled steam-powered machinery spread rapidly across Great Britain, transforming the way goods were manufactured.

Information molecule: a general name for a molecule that encodes information.

Invertebrates: animals that are not vertebrates.

Ion: a charged atom that has gained or lost electrons.

Junk DNA: non-coding DNA that has no apparent use.

Kelp: a large brown seaweed that can form underwater forests.

Keystone species: a species that has far-reaching impacts on the ecosystem it inhabits.

Kidneys: a paired organ in vertebrates that excretes nitrogen-containing waste products.

Krill: a small shrimp-like animal that occurs throughout the world but is particularly abundant in the Southern Ocean.

Lactic acid: a toxic substance produced by muscles when they don't have enough oxygen.

Lactose: a type of sugar.

Larva: the pre-adult stage of many animals.

Legumes: a group of plants including peas, beans and lentils that are high in nitrogen.

Lichens: crusty or leafy organisms that can be found growing on tree trunks or rocks. Can be mistaken for plants but consist of a fungus and at least one species of alga that live and work together.

Life-history trait: a feature of an organism, such as the number of offspring, that is closely linked to its population growth.

Lignin: a molecule found in wood that provides strength and is very difficult for other organisms to digest.

Lipid: a type of fatty molecule from which cell membranes are made.

Litter: a group of offspring that are born at the same time. Usually applied only to mammals.

Liver: an organ within vertebrate bodies that detoxifies molecules, including surplus amino acids.

Loki's Castle: a system of deep-sea hydrothermal vents in the mid Atlantic Ocean.

Lung: an organ that allows gas exchange between the air and the blood.

Lungfish: a type of fish that possesses a simple lung and can survive out of water for a long period of time.

Lycophytes: a type of plant that formed trees during the Palaeozoic.

Lystrosaurus: an extinct animal that survived the Great Dying at the end of the Palaeozoic and spread around the world.

Macromolecules: a large molecule that is made by joining together smaller sub-units.

Macrophage: a type of white blood cell found in mammalian immune systems that crawls around and digests invading cells.

Maize: an important crop domesticated by humans in South America.

Malaria: a serious disease caused by a parasite that is transmitted by mosquitoes.

Marsupial: a type of mammal in which the baby animal develops and grows outside its mother's body in a pouch.

Mass extinction: an event that is rapid in geological terms and results in the extinction of around three-quarters of the species present at the time. Five are generally recognized.

Megafauna: a set of very large animals that dominated during the Pleistocene and went extinct around the time that the last ice age ended.

Mesozoic: the middle era of life on Earth. Runs from 252 million years ago to 65 million years ago.

Metabolic rate: the rate of energy expenditure per unit time.

Microbiome: the community of bacteria and archaea living harmlessly inside other organisms, often in their guts where they help to digest food.

Mitochondria: the powerhouses of eukaryotic cells that were once free-living bacteria.

Moa: large flightless birds that used to live in New Zealand but are now extinct.

Molecule: a collection of atoms that are bonded together by shared electrons.

Mollusc: a phylum within the Animal Kingdom that contains snails, bivalves and cephalopods (octopus and squid).

Monotreme: a type of mammal that lays eggs rather than bearing live young.

Mosquito: a small insect that sucks the blood of vertebrates, including humans.

Motor neurons: nerve cells that carry messages from the central nerve cord or brain to the muscles.

Muscle cells: a cell that can change its length and so cause the movement of body parts.

Mutagen: a substance or chemical that makes it more likely that genetic changes will occur.

Mutation: a permanent change to the letters in the genome that can be passed from parent to offspring.

Mutualism: a relationship between two different organisms in which both gain benefits.

Mycorrhizae: a type of soil-dwelling fungus that has a mutualistic relationship with plants.

Natural selection: the process that causes beneficial genetic changes that arise by chance in a single individual to spread through a population.

Nectar: a sugary solution produced by flowers to encourage insects to visit.

Negative feedback: a process whereby an abundance of a molecule inhibits its own production, so that it is only manufactured when scarce.

Nematode: a type of worm.

Neolithic: New Stone Age. Lasted from roughly 12,000 to 6,500 years ago, during which time humans began to farm and form settled communities.

Nerve cell: a cell that conducts electrical signals and so allows animals to make co-ordinated movements and respond to their environment.

Nerve cord: a bundle of nerve cells that runs along the back of bilaterian animals.

Neurons: nerve cells.

Neurotransmitters: chemical signals secreted by nerve cells so that they can communicate with other cells, such as muscle cells.

Neutron: a particle found in the nuclei of atoms that does not carry an electrical charge.

Niche: the ecological requirements of an organism that allow it to thrive.

Nitrogen-fixing bacteria: bacteria that live inside the roots of leguminous plants and contain unique enzymes that convert nitrogen gas from the air into ammonia (NH_3).

Non-coding DNA: parts of the genome that do not code for proteins.

Nucleotides: the building blocks of the information molecules DNA and RNA. Often referred to by their initial letters.

Nucleus (atomic): the centre of an atom that contains protons and neutrons.

Offspring: children.

Opportunist pathogen: an organism that is mostly harmless, but that can cause disease if the opportunity arises.

Organ: a group of cells that work together to perform a specific function within an animal's body.

Organelles: membrane-bound compartments of different shapes and sizes and with different functions that are found inside eukaryotic cells.

Organic molecules: molecules that are based on carbon and are often made by living things.

Osmosis: the movement of water across a membrane from a dilute solution to a more concentrated one – a special case of diffusion.

Ozone layer: a layer in the Earth's atmosphere around 15 to 35 km above the surface that contains short-lived molecules of ozone (O_3). Ozone absorbs ultraviolet (UV) radiation that would otherwise damage DNA.

Palaeolithic: Old Stone Age. Runs from the first emergence of modern humans until roughly 12,000 years ago.

Palaeozoic: the era of ancient life on Earth. Runs from 541 million years ago to 252 million years ago.

Paramecium: a single-celled free-living eukaryotic cell that lives in ponds.

Parasite: an organism that lives on or in another individual, stealing resources for its own growth and reproduction.

Parasitoid: organisms whose larvae live a parasitic lifestyle, such as members of the wasp family that lay their eggs inside other host species.

Pathogen: an organism (usually a virus or a bacteria) that lives inside another organism (often called the host) and that causes disease.

Penicillin: the first antibiotic.

Period: a slice of geological time, shorter than an eon or an era. (See also Eon and Era).

pH: a measure of acidity.

Phloem: narrow pipes inside plants that transport sugar.

Photosynthesis: the process of converting carbon dioxide and water into glucose and oxygen using energy from the sun.

Photosystem: a huge molecular machine invented by bacteria that uses the energy in sunlight to fire off electrons by destroying other molecules, like water.

Phylum/Phyla: a high-level grouping within the Animal Kingdom.

Placenta: an organ that provides oxygen and nutrients to a baby mammal inside its mothers body.

Placental: a type of mammal in which the foetus develops and grows inside its mother's body, nourished by its mother's blood.

Plant: a green organism descended from algae that conducts photosynthesis.

Plasmid: a small piece of DNA that is separate from the chromosome and that can be passed among cells. Found in bacteria and archaea.

Pleistocene: a period of time during the Cenozoic in which ice ages gripped the Earth.

Plesiosaur: a large extinct marine reptile with a round body, four paddle-like limbs and a long thin neck.

Pluripotent: a cell that still has the potential to become many different types of cell within an animal's body.

Pneumonia: a bacterial infection of the lungs.

Primary producers: any organism (like plants or algae) that turns carbon dioxide into sugar.

Prokaryotes: a collective name for bacteria and archaea.

Promoter: a special sequence of DNA letters at the beginning of a gene that helps to bind the molecular machine that produces the RNA message.

Protein: large molecules made from amino acids that can be strung together in any order.

Protein synthesis: the process of manufacturing proteins inside cells.

Proterosuchus: an extinct crocodile-like animal that probably preyed on *Lystrosaurus*.

Proton: the smallest unit that carries a positive charge. Found in the nuclei of atoms together with neutrons.

Pseudopodia: also known as 'false feet'. Bulging blobs of cytoplasm that push out from some large eukaryotic cells, such as amoeba, and allow the cell to crawl around.

Pterosaur: an extinct winged reptile that flew using wings made from a thin skin membrane supported by an enormously long bony finger.

Quorum sensing: a method used by cells to communicate and estimate their own numbers.

Radula: a highly modified tongue used by snails to scrape food from hard surfaces.

Recombination: the process of cutting and pasting the two slightly different halves of the double genome carried by all sexual species to make new combinations of genes.

Repressor protein: a protein that binds to a promoter sequence and turns off a gene.

Reptile: the group of tetrapods that includes modern lizards, turtles, crocodiles and birds, as well as other extinct groups, such as pterosaurs, ichthyosaurs and dinosaurs.

Ribosome: a large molecular machine that builds proteins by stringing together amino acids.

RNA (ribonucleic acid): an information molecule that consists of any number of RNA nucleotides strung together to form a long strand. Messenger RNA carries messages from the genome to the ribosomes. Ribosomal RNA is found inside ribosomes.

Ruminants: a group of mammals, including cows and sheep, that have multiple stomachs to digest their food. One stomach is called the rumen, and it contains bacteria that digest cellulose.

Rust: iron oxide.

Sauropod: a group of enormous herbivorous dinosaurs with four columnar legs, and a long neck and tail. Examples include *Diplodocus* and *Brachiosaurus*.

Sea squirts: a group of marine animals that are closely related to vertebrates and belong to the same phylum.

Seaweed: multicellular algae.

Second law of thermodynamics: a physical law demanding that the entropy of the universe must continually increase.

Seed plants: plants that produce seeds.

Selfish genetic elements: regions of DNA that copy and insert themselves into other parts of the genome.

Sensory neurons: nerve cells that carry messages from sense organs, like eyes, to the central nerve cord or brain.

Sex chromosomes: only present in some species, like mammals and birds, this pair of chromosomes determine the sex of offspring.

Sexual reproduction: the production of offspring through mixing genomes from two different parents.

Sexual selection: a type of natural selection that acts on features of an individual that allow it to attract mates.

Sickle cell disease: a genetic disorder that affects the oxygen-carrying protein haemoglobin.

Slime molds: amoeba-like organisms that gang up to form larger structures that move around.

Specialist: an organism that eats only one type of food.

Sperm cell: a male sex cell that is usually small and mobile.

Spliceosome: a massive piece of machinery inside the nucleus of eukaryotic cells that performs a cut-and-paste job on RNA messages by removing stretches of nonsense.

Sponges: a phylum within the Animal Kingdom. The adults have a simple body plan and are immobile filter feeders, while the larvae can swim.

Spontaneous reaction: a chemical reaction that is naturally favoured. Very often energy is released.

Spore (bacterial): a long-lived structure with a resistant outer wall that can give rise to a new bacterial cell and can last for thousands of years. It allows certain bacteria to sit out difficult times and reactivates when conditions improve.

Stamen: the male part of a flower that produces pollen.

Stigma: the uppermost part of the carpel – the female part of the flower – where pollen lands.

Stomata: tiny holes in the cuticle of a plant leaf that allow gases and water to enter and leave.

Stromatolite: a mound of hard sediment found in shallow waters that has been constructed by cyanobacteria. They have been around for the last 3.5 billion years (although some may have been made by other types of bacteria).

Superbug: any type of pathogenic bacteria that is resistant to multiple different antibiotics.

Surface area to volume ratio: a crucial aspect of animal size.

Teosinte: the wild ancestor of maize.

Termite: an insect that lives in organized societies where individuals play different roles.

Tetrapods: vertebrates with four limbs.

Therapod: a type of bipedal dinosaur that gave rise to the birds. Examples include *Velociraptor* and *Tyrannosaurus rex*.

Through-gut: a gut that has two openings, a mouth and an anus, so that food moves through in one direction.

Totipotent: a cell that still has the potential to become any cell within an animal's body.

Trilobite: an extinct group of arthropods that were very common during the Palaeozoic.

Turgor pressure: the pressure exerted by water molecules pushing outwards against the cell wall that provides strength and support to plants.

Ultraviolet radiation: a high-energy type of radiation emitted by the sun.

Urchin barrens: areas of seabed scraped clean of all algae through overgrazing by sea urchins.

Urea: a waste product containing nitrogen that is produced in the livers of mammals from surplus amino acids.

Uric acid: a waste product containing nitrogen that is produced in the livers of reptiles (including birds) from surplus amino acids.

Urine: a liquid excreted by the kidneys of mammals that contains the nitrogen-containing waste product, urea.

Virulence genes: genes possessed by opportunistic pathogens that can be switched on when inside a host to maximize damage. These genes are often expressed in a co-ordinated way.

Virus: an entity that carries a genome within a protein coat. It is incapable of reproducing without hijacking the machinery of living cells.

Wandering albatross: a seabird with the longest wingspan of any bird.

Xylem: pipes inside plants that transport water from the soil to the leaves.

THANKS

This book is the product of a lifetime spent being crazy about biology. As such, there are a lot of people to thank. First, I want to thank my parents who encouraged and supported my interest – I still have the pair of binoculars they bought me when I was eight years old, and my copy of *The Spotter's Guide to Wildlife* with which I identified the weeds in the pavements around my house. Many years later, after completing a PhD and spending 10 years in Switzerland, I arrived at the University of Oxford and joined the Department of Plant Sciences. Finding myself surrounded by wonderful colleagues both there and in the Department of Zoology (now combined into a new Biology Department), I decided to try to write this book.

I showed my first efforts to Stuart West, who encouraged me to continue and introduced me to David Fickling at DFB. David has been a constant source of enthusiasm and campaigned tirelessly to improve my writing – I owe him a huge debt of thanks. Polly Shields helped me to understand what metaphors are and how to use (and misuse) them – not to mention always being a fantastic friend. Thanks also to Frances Sheahan (another fantastic friend), who continually reminded me that plenty of people would like to understand more about biology.

I received inspiration from colleagues in Seychelles through my association with the Seychelles Islands Foundation – thank you in particular to Frauke, Nancy, April, Jude, Jeremy and Maurice Loustou-Lalane for sharing your knowledge of Aldabra and allowing me to visit. Thanks also to the Family Sigrist (Petra, Roland, Jakob and Lena) and their fabulous house in Obertschappina, and to Tim Smit for inspiring me not to be afraid to follow my own ideas.

Once the chapters started to come together, I was fortunate enough to find colleagues who were prepared to help find the talking points and give comments. In particular, I would like to thank Lee Sweetlove for his incomparable knowledge of biochemistry, cell and molecular biology; Ashleigh Griffin for her insights into evolution; Kayla King for her expertise on sexual reproduction; Seb Shimeld for being dedicated to all those invertebrates that most of us ignore; Peter Holland for knowing everything about everything, but especially for knowing about vertebrates and their development; and Sandy Hetherington for his outstanding knowledge of plants both past and present. None of these people bear any responsibility for any outstanding errors or mistakes – no doubt they tried to warn me, but I had to make sometimes difficult calls about how much detail to leave in. Many other colleagues also enlightened me about a whole range of topics during innumerable conversations over coffee, particularly Phil Poole, Stu West, Steve Kelly and Tamsin Mather – thank you all for sharing your knowledge so generously.

One of the greatest joys of my life has been to teach biology students at The Queen's College, Oxford. The Oxford tutorial system, while highly demanding of staff time, is a wonderful throwback to a world where people really prized the personal touch. Teaching enthusiastic students in small groups is a real privilege, and engaging with my students helped me understand what needed to be in this book and what could be left out. My heartfelt thanks to them all and to my colleagues at The Queen's College who continue to uphold the tutorial tradition, while at the same time being great fun and wonderful to work with.

Once the chapters were drafted, they benefited enormously from the input of my editors, Rosie and David Fickling, who acted the part of non-biologist readers with great aplomb and helped me get to the heart of every subject. Julia Bruce also did a great job of the copy-edit, and picked up more than just grammatical errors. In the final phase, I worked with Cécile Girardin on the illustrations, which was a wonderful experience. I would draft something that mostly looked awful and she would turn it into something beautiful and amazing. Katie Bennett also worked tirelessly on the book design, constructing the layouts and adding labels to the figures, with the help of Rosie's eagle eyes. Finally, I had to write this book at weekends, evenings and on holidays. I used to hate reading the acknowledgements in books written by male scientists who always thanked their long-suffering wives for having basically put up with them not being there. I hope I'm not guilty of the same, and that my partner Andy and son Rowan don't think that I ever abandoned them or got the work – life balance badly wrong. Rowan – this book was always for you and thank you for your helpful contributions. I'm sorry it took so long to finish, but hope it's still useful. Andy – thank you for being so supportive of my career.

Lindsay Turnbull is Professor of Plant Ecology at the University of Oxford and a Tutorial Fellow at The Queen's College, Oxford. She loves teaching and has been a secondary-school science teacher as well as Director of Undergraduate Biology Teaching. She has published scientific papers on a wide range of topics, including plant ecology, interactions between legume plants and their nitrogen-fixing bacteria and plastic pollution on remote islands. Her favourite animal is (probably) the sea hare.

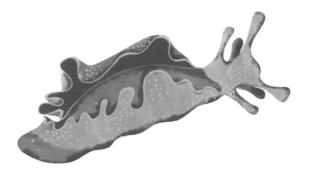

Dr. Cécile Girardin is a scientist and an artist. As an artist, she is passionate about illustrating science stories. Cécile has produced murals, animations and a wide range of illustrations for UNFCCC, UN Biodiversity, IIED, Oxford University Press, the Global Canopy Programme, the Sumatran Orangutan Society, the British Ecological Society, and David Fickling Books, to name a few. Her portfolio is available on @cecilegirardin

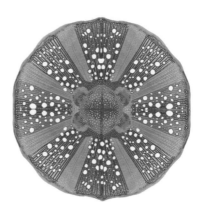

FURTHER READING

The major dates for events in the history of life are drawn from:

Betts et al. 2018: Integrated genomic and fossil evidence illuminates life's early evolution and eukaryote origins. *Nature Ecology and Evolution* 2: 1556 –1562

The following textbooks are excellent and authoritative sources of information, although some provide a level of detail that is far beyond this book.

Molecular Biology of the Cell. Alberts *et al*. (Garland Science, New York)

Brock Biology of Microorganisms. Brock *et al*. (Pearson)

Cell Biology by the Numbers. Philips and Milo (Garland Science, New York)

Vertebrate Life. F Harvey Pough (Pearson)

Botany. James Mauseth (Jones & Bartlett Publishers, Inc)

The following *Very Short Introductions* (OUP) are shorter and more accessible.

The Animal Kingdom by Peter Holland

The History of Life by Michael J Benton

Earth System Science by Tim Lenton

Bacteria by Sebastian G. B. Amyes

Molecules by Philip Ball

Sexual Selection by Marlene Zuk and Leigh W. Simmons